URBAN HABITATS

Conjuring up images of disused canals filled with old shopping trolleys, or derelict building plots infested with fly-tipped refuse, habitats in urban areas are commonly thought to be of little value to nature conservation. Yet our towns and cities contain sites which feature a wide variety of plants and animals. Some of these habitats are remnants of rural sites encompassed by urban sprawl, while others are uniquely urban in nature. Irrespective of their origin, many urban habitats provide interesting places for nature-watching and serious ecological study.

Urban Habitats presents an illustrated and practical guide to the wide range of urban habitats and the flora and fauna which live within them, and examines the most important conservation and management issues faced within our towns and cities.

Following an explanation of the nature of urban habitats, the range of habitat types and typical plants and animals are described in detail. Topics of concern to the conservationist or habitat manager are explored: the impacts and monitoring of pollution; the effects of invasive species; and guidelines for the ecological management of sites to enhance their nature conservation value. A series of case studies illustrate the pressures on some major urban habitat types, and the opportunities to enhance them for nature conservation.

Featuring up-to-date references and data, illustrated species boxes with typical plants and animals, and a series of projects on the ecology, behaviour and distribution of urban species, students, environmentalists and the urban dweller will gain a deeper understanding of the nature of the rapidly expanding urban environment through this illuminating habitat guidebook.

C. Philip Wheater is Principal Lecturer in Ecology, Department of Environment and Geographical Sciences, Manchester Metropolitan University.

HABITAT GUIDES
Series editor: **C. Philip Wheater**

Other titles in the series:

Upland Habitats
Woodland Habitats

Forthcoming titles:

Freshwater Habitats
Grassland and Heathland Habitats
Marine Habitats
Agricultural Habitats

URBAN HABITATS

●

C. Philip Wheater

Illustrations by Jo Wright

LONDON AND NEW YORK

First published 1999
by Routledge
11 New Fetter Lane, London EC4P 4EE

Simultaneously published in the USA and Canada
by Routledge
29 West 35th Street, New York, NY 10001

Typeset in Sabon by RefineCatch Limited, Bungay, Suffolk
Printed and bound in Great Britain by
The Bath Press, Bath

British Library Cataloguing in Publication Data
A catalogue record for this book is available from the British Library

Library of Congress Cataloging in Publication Data
Wheater, C. Philip, 1956–
Urban habitats / C. Philip Wheater.
p.   cm. – (Habitat guides)
Includes bibliographical references (p.   ).
1. Urban ecology (Biology)   2. Nature conservation.   I. Title.
II. Series.
QH541.5.C6W465   1999
577.5′6 – dc21   98–34690

ISBN 0–415–16264–5 (hbk)
ISBN 0–415–16265–3 (pbk)

CONTENTS

●

PLATES

●

FIGURES

●

TABLES

•

SPECIES BOXES

●

ACKNOWLEDGEMENTS

●

My early interest in ecology was encouraged by my family and later stimulated and moulded by a number of people, mostly at the Zoology Department (later the Department of Environmental Biology) at The Victoria University of Manchester. I am especially indebted to Robin Baker, Gordon Blower, Glyn Evans, Mike Hounsome, Martin Jones and Derek Yalden.

Many people helped in the production of this book, some provided access to material and sites, and gave their time to discuss issues and case studies. Others read, commented on and discussed sections of the text. All responded to requests for help with enthusiasm. I offer grateful thanks to: Kate Ainsworth (British Waterways); Roger Cook (Central Science Laboratories); Roy Croucher (Sandwell Valley); Hugh Firman; Chris Gibson (English Nature); David Holland (The Environment Agency); Mike Hounsome (Manchester Museum); Andrew Littlewood (Camley Street Natural Park); Pauline Mousdell (St Helens, Knowsley and Sefton Groundwork Trust); Dave Needham (Tunstead Quarry, Buxton Stone); Chris Powell (Cardiff County Council); Helen Read (Corporation of London); Richard Scott (Landlife, Liverpool); Robert Wardell (Cardiff County Council); and Nina West (Mersey Forest).

I am also grateful to staff at the Countryside Commission; the Groundwork Foundation; Royal Society for the Protection of Birds; as well as colleagues from Stockholm University (Arja Kaitala and Nina Wedell) and the Manchester Metropolitan University (James Bell, Paul Chipman, Rod Cullen, Mike Dobson, Alan Fielding, Mark Langan, Gregg Paget, Liz Price and Amanda Wright).

I would also like to thank Liz Potts for helping to obtain much of the literature, Jo Wright for producing the figures and line drawings, Sarah Lloyd for the initial idea and editorial help along the way, several referees whose comments helped to clarify some ambiguities, and Penny Cook for reading, commenting on, and amending several versions of the text. I am grateful to Professors Steve Dalton and Christer Wiklund for the facilities provided by the Department of Environmental and Geographical Sciences at the Manchester Metropolitan University and the Zoology Department at Stockholm University where most of the writing took place.

To all of these I offer my thanks. Without their help, writing this volume would have been much more difficult and the resulting text less complete, although any errors which remain are mine alone. My overwhelming thanks go to Penny Cook who has supported me in this venture from the beginning, and without whom it would not have been possible.

SERIES INTRODUCTION

•

The British landscape is semi-natural at best, having been influenced by human activities since the Mesolithic (*circa* 10,000–4,500 BC). Although these influences are most obvious in urban, agricultural and forestry sites, there has been a major impact on those areas we consider to be our most natural. For example, upland moorland in northern England was covered by wild woodland during Mesolithic times, and at least some was cleared before the Bronze Age (*circa* 2,000–500 BC), possibly to extend pasture land. The remnants of primaeval forests surviving today have been heavily influenced by their usage over the centuries, and subsequent management as wood-pasture and coppice. Even unimproved grassland has been grazed for hundreds of years by rabbits introduced, probably deliberately by the Normans, sometime during the twelfth century.

More recent human activity has resulted in the loss of huge areas of a wide range of habitats. Recent government statistics record a 20 per cent reduction in moorland and a 40 per cent loss of unimproved grassland between 1940 and 1970 (Brown, 1992). In the forty years before 1990 we lost 95 per cent of flower-rich meadows, 60 per cent of lowland heath, 50 per cent of lowland fens and ancient woodland, and our annual loss of hedgerows is about 7,000 km. There has been substantial infilling of ponds, increased levels of afforestation and freshwater pollution, and associated reductions in the populations of some species,

especially rarer ones. These losses result from various impacts: habitat removal due to urban, industrial, agricultural or forestry development; extreme damage such as pollution, fire, drainage and erosion (some or all of which are due to human activities); and other types of disturbance which, although less extreme, may still eradicate vulnerable communities. All of these impacts are associated with localised extinctions of some species, and lead to the development of very different communities to those originally present. During the twentieth century over one hundred species are thought to have become extinct in Britain, including 7 per cent of dragonfly species, 5 per cent of butterfly species and 2 per cent of mammal and fish species. Knowledge of the habitats present in Britain helps us to put these impacts into context and provide a basis for conservation and management.

A habitat is a locality inhabited by living organisms. Habitats are characterised by their physical and biological properties, providing conditions and resources which enable organisms to survive, grow and reproduce. This series of guides covers the range of habitats in Britain, giving an overview of the extent, ecology, fauna, flora, conservation and management issues of specific habitat types. We separate British habitats into seven major types and many more minor divisions. However, do not be misled into thinking that the natural world is easy to place into pigeon holes. Although these are convenient divisions, it is important to recognise that there is considerable commonality between the major habitat types which form the basis of the volumes in this series. Alkali waste tips in urban areas provide similar conditions to calcareous grasslands, lowland heathland requires similar management regimes to some heather moorland, and both estuarine and lake habitats may suffer from similar problems of accretion of sediment. In contrast, within each of the habitat types discussed in individual volumes, there may be great differences: rocky and sandy shores, deciduous and coniferous woodlands, calcareous and acid grasslands are all typified by different plants and animals exposed to different environmental conditions. It is important not to become restricted in our appreciation of the similarities which exist between apparently very different habitat types and the, often great, differences between superficially similar habitats.

The series covers the whole of Britain, a large geographical range across which plant and animal communities differ, from north to south and east to west. The climate, especially in temperature range and precipitation, varies throughout Britain. The south-east tends to experience a continental type of climate with a large annual temperature range and maximal rainfall in the summer months. The west is influenced by the sea and has a more oceanic climate, with a small annual temperature range and precipitation linked to cyclonic activity. Mean annual rainfall tends to increase both from south to north and with increased elevation. Increased altitude and latitude are associated with a decrease in the length of the growing season. Such climatic variation supports different species to differing extents. For example, the small-leaved lime, a species which is thermophilic (adapted for high temperatures), is found mainly in the south and east, while the cloudberry, which requires lower temperatures, is most frequent on high moorland in the north of England and Scotland. Equivalent situations occur in animals. It is, therefore, not surprising that habitats of the same basic type (such as woodland) will differ in their composition depending upon their geographical location.

In the series we aim to provide a comprehensive approach to the examination of British habitats, whilst increasing the accessibility of such information to those who are interested in a subset of the British fauna and flora. Although the series comprises volumes covering seven broad habitat types, each text is self-contained. However, we remind the reader that the plants and animals discussed in each volume are not unique to, or even necessarily dominant in, the particular habitats but are used to illustrate important features of the habitat under consideration. The use of scientific names for organisms reduces the likelihood of confusing one species with another. However, because several groups (especially birds and to a lesser extent flowering plants) are often referred to by common names (and for brevity), we use common names where possible in the text. We have tried to use standard names, following a recent authority for each taxonomic group (see the species list for further details). Where common names are not available (or are confusing), the scientific name has been used. In all cases, species mentioned in the text are listed in alphabetical order in the species index and, together with the scientific name, in systematic order in the species list.

1

INTRODUCTION

•

Urban habitats sometimes conjure up images of disused canals filled with old prams and shopping trolleys, or derelict building plots gradually accumulating a cover of vegetation amongst a mass of fly-tipped refuse. However, an amazing range of habitats with their associated plants and animals are found in towns, cities and areas of urbanisation on the urban fringe. Including terrestrial and aquatic areas, these range from semi-natural sites that were enclosed by the spread of towns and cities, to artificial habitats formed during urban developments. Semi-natural sites include old woodlands, heathlands, parklands, river valleys and wetlands whilst other sites such as meadows, small woodlands, ponds, hedges and ditches may be remnants of agricultural land. Many open spaces within the town or city are a result of urban residence (for example parks and gardens, playing fields, buildings, cemeteries and sewage works) or are associated with disuse and decay (such as abandoned industrial land, waste tips and quarries).

Urban habitats are those within the confines of a town or city. The definition of a habitat on the urban fringe is less straightforward. The urban fringe can be considered as the boundary between urban areas and the wider countryside, particularly where otherwise rural areas have been subject to the impact of urbanisation. I use the term 'urban fringe' to encompass habitats such as derelict industrial land and mineral workings that are essentially urban but lie on the edge of conurbations or intrude into otherwise rural areas. The aim of this volume is to examine the range of habitats and their associated fauna and flora in urban and urban fringe areas, discuss management and conservation issues and suggest ways of investigating and monitoring urban ecology. It is perhaps unsurprising that habitats tend to be defined by the plants which live there. Plants are usually the most obvious residents in a site, being in many ways the structural components; providing shelter and food for many animals. In this book I discuss both the flora and fauna, although to avoid repetition some species are dealt with in only one of their major habitats. Urban habitats include terrestrial and aquatic sites, both inland and at the coast. Since many habitats in urban areas are covered in more depth by other volumes in the series, this book concentrates on those terrestrial and, to a lesser extent, aquatic habitats in towns and cities which are typical of urban areas.

As towns and cities develop, rural landscapes are destroyed or engulfed and new urban habitats are created. The rise in the world's human population, from about 5.2 billion in 1990 to 8 billion or so in 2025, will increase the proportion living in urban areas (from 34 per cent in 1960 to 44 per cent in 1990 and probably about 60 per cent by 2025). This will not only bring more areas under the urban umbrella, but will also influence existing habitats in towns and cities. By

1981, 10.2 per cent of land in England was in urban use. This is expected to rise to 11 per cent by 2001, with more in regions such as the south-east (Brown, 1992). However, the development of urban areas was much more rapid over the period 1951–71 (Davis, 1976), with an overall increase for the United Kingdom of 22 per cent (27 per cent for England, 33 per cent for Wales and 5 per cent for both Scotland and Northern Ireland). This reduction in rate of increase is due, at least in part, to a recent rise in the use of previously developed land for new building programmes (residential, commercial and industrial).

Despite more efficient land use, substantial areas of vacant urban land still exist throughout the country (see Table 1.1 for details). In 1990, approximately 5 per cent of the land area of large English towns was vacant, about 43 per cent of which had once been developed. With the addition of smaller towns, there is about 60,000 ha of vacant urban land in England, 25,000 ha of which had previously been developed. Recycling vacant land is obviously preferable to developing rural land. However, some urban land has been vacant for a considerable time; in Scotland, 44 per cent was vacant for at least ten years. These time-scales

Table 1.1:   Urban vacant land distribution in England and Scotland in areas with the highest and lowest amounts of derelict land (calculated from towns with more than 10,000 inhabitants)

Region		Vacant land as per cent of urban area
England (surveyed in 1990) (approximately 49,000 ha)		5
Highest	Northern Region	11
	Yorkshire and Humberside	6.8
	North West	6.7
	East Midlands	6.2
Lowest	South West	3.7
	South East	3.3
	Greater London	2.8
Scotland (surveyed in 1988) (approximately 5,060 ha)		2.2
Highest	Strathclyde	4
	Central	3
	Lothian	2.3
Lowest	Borders	<1
	Orkney	<1
	Dumfries & Galloway	<0.5
	Western Isles	<0.5

Source: Brown (1992)

lead to the establishment of interesting floras and faunas which are disturbed or destroyed by redevelopment.

Urban landscapes are dominated by developments: buildings (residential and industrial); transport links (roads, pavements, railways, canals); and open land required for parking vehicles and disposal of waste. There are also, sometimes substantial, areas of greenspace: in Leicester, gardens, parks, playing fields, golf courses, road verges and grounds of public institutions comprise nearly half of the land area (Baines, 1995). Land at the centre of urban areas is usually in demand and lacks extensive greenspace; that which is there is often planted and designated as parks. Further from the centre, residential and industrial pressures tend to ease and there is an associated increase in greenspace, often retained for historical reasons, or resulting from the colonisation of abandoned areas or deliberate planting of vegetation. Urban areas are not static; they are constantly expanding around the edges whilst inner regions degenerate, some to be abandoned for a while and others to be redeveloped. This dynamic system influences and explains the development of urban habitats.

Industrial growth is usually associated with a rise in the extent of derelict and waste land, as factories close either because of decline or transfer elsewhere. The resulting habitats are heavily influenced by pollution, disturbance or fragmentation, all of which impact on the plant and animal communities. Even relatively untouched sites which were encapsulated by the growth of towns and cities are affected by the removal of direct links with the surrounding countryside. There has been a recent increase in reclamation, restoration and habitat creation in an attempt to improve urban sites. The resulting habitats are artificial, but mimic natural environments. Other artificial

habitats are accidental analogues of natural habitats. The sheer faces of buildings and quarries act as cliffs for nesting and roosting birds; street lamps, telegraph poles and pylons provide perching sites; cellars and roof spaces substitute as caves for bats; subsidence in flooded mine workings produce open water and marshland.

Urban areas house many species. Up to sixty species of birds have been recorded from cities, with between twenty and thirty from large city centres and up to thirty-four breeding species in city centre parks (Hounsome, 1979). Although large and less tolerant mammals, such as deer and hares, avoid heavily industrialised areas, many mammals live in suburban habitats and sites such as railway embankments and motorway verges. These include small mammals such as wood mice and shrews as well as badgers and foxes. Urban habitats may support native and introduced plants, and sometimes include orchids. Over 240 species of fungi occur in the London Borough of Haringey. In this urbanised area, many common rural species were found, especially in woodlands, cemeteries and along railways (Thomas, 1992). Thus, although species in urban sites are not usually unique, they may form interesting communities which exploit the special conditions in these disturbed environments. Novel combinations of characteristics lead to some urban sites having a high nature conservation value: in 1990 nearly 30 per cent of all Sites of Special Scientific Interest (SSSI) in England were classed as urban or urban fringe (Eggo, 1990). Increasingly, newly designated SSSIs within urban areas are not previously rural sites surrounded by urban sprawl, but are sites which, through a novel combination of environmental features, are important on a regional or national basis. The physical, ecological and geographical make up of urban sites influence the species which may be found. These factors are discussed below.

IMPACTS OF PHYSICAL FACTORS

Soil quality determines plant growth, being involved in nutrient absorption and supply, water holding and as a habitat for soil organisms. Urban soils are often shallow, polluted, or deficient in some quality, for example lacking structure, organic material, nutrients or water. Comparisons between urban and more natural soils are shown in Table 1.2. The supply of nutrients and water influences which

pioneer species establish on newly created sites (Ash, 1991: see Table 1.3 for examples). Organic matter also enhances the fixation of organic pollutants and generally improves soil ecology. Many urban soils would be improved by stimulation of humus formation (partially decomposed organic matter), which is enhanced by the retention of litter material, avoidance of herbicides, and by applying extra organic material (Beyer *et al.*, 1995).

Urbanisation affects the local climate. Urban areas tend to be less windy, have a higher rainfall, and be warmer and less humid than nearby rural areas (Chandler, 1976). City winds are more turbulent due to air moving

Table 1.2: General differences between urban and more natural soils

Characteristics of urban soils in comparison with natural soils	Causes	Resulting problems
Harsh boundaries between soil layers	Artificial origins produce layering of different materials	Lack of continuity for rooting plants and burrowing soil animals
Compaction	Trampling and pressure from vehicles	Reduced water passage and lack of air spaces. Plants produce shallow roots
High pH	Effects of de-icing salts, and water running over calcareous building materials (e.g. concrete)	Problematic if highly alkaline, because some nutrients (e.g. phosphates) are immobilised.
Low water drainage	Diversion of run-off to drains, and interruption of natural flow through soils	Reduced water availability for plants
Restricted nutrient cycling	Low levels of organic material, lack of organic input from above ground communities and reduced soil organism communities	Reduced water-holding capacity and nutrient status
Pollution	Incorporation of contaminants in soil layers and polluted run-off	Toxic effects (lethal or sublethal) on plants and animals
High soil temperatures	Higher ambient air temperatures and little buffering effect of vegetation	Reduced moisture in upper layers for plant growth

Sources: Bradshaw and Chadwick (1980); Harris *et al.* (1996)

Table 1.3: Examples of pioneer plants in urban sites

Examples of pioneer plant species	Nutrient level	Water level	Example habitats
Herbs (bur-marigold species, knotgrass species, red goosefoot, spear-leaved orache)	★★★★	★★★★	Sewage works
Grasses (barren brome, wall barley) Herbs (common chickweed, fat-hen, groundsel, speedwell species)	★★★★	★★★	Gardens
Grasses (common bent) Herbs (charlock, colt's-foot, dock species, Oxford ragwort, rocket species, scentless mayweed, wall-rocket species)	★★	★★	Demolition sites
Herbs (corn spurrey, scarlet pimpernel, speedwell species)	★★	★★	Railway tracks
Liverworts (*Marchantia polymorpha*) Mosses	★	★★★	Viaducts
Mosses (purple-fruiting heath-moss, silvery thread-moss)	★	★	Tarmac
Liverworts (*Lunularia cruciata*) Mosses (common cord-moss, *Leptobryum pyriforme*)	★	★	Brick
Lichens (*Xanthoria* species, *Candelariella* species) Mosses (Beard moss species, *Bryum* species, Wall-screw moss)	★	★	Concrete

Abundant ★★★★   Adequate ★★★   Limited ★★   Deficient ★

Source: Ash (1991)

around buildings, and are channelled and accelerated down streets orientated along the same direction as the wind. The turbulence associated with cities increases cloud cover and rainfall, although drainage from impervious urban surfaces hastens the removal of standing water, thus reducing the humidity. Higher temperatures also contribute to lower humidities. A mass of warm air often covers urban areas, producing a heat island, usually 1–2°C, and sometimes as much as 5–10°C, warmer than surrounding rural temperatures. Larger cities buffer the effects of winds to a greater extent and high building densities promote sharp boundaries between the warmer cities and cooler rural areas, especially at night.

These urban climatic conditions influence plants and animals, both directly and via impacts on atmospheric pollution. For example, the heat-island effect reduces the likelihood of ground and air frosts in cities and increases the active plant growing season. In London, the active growing season is 279 days in Regent's Park compared to 264 days in the suburbs and 259 days in nearby rural Surrey (Davis, 1982a). The success of Oxford ragwort in urban sites may be due to the warmer winter conditions found in cities (Grime *et al.*, 1988). There are also impacts on animals;

snails in the centre of London, where conditions are drier, tend to be larger than at the edge (Cousins, 1982). This may be because large snails have a lower surface area to volume ratio and hence a lower rate of water loss. Increased heat in built-up areas has been implicated in the localised extinction of the snail, *Arianta arbustorum*, in Basel, Switzerland, possibly because its eggs are intolerant of high temperatures (Baur and Baur, 1993).

ECOLOGICAL ASPECTS

Colonisation and Establishment

Communities do not simply depend on physical factors, they are heavily influenced by movements between neighbouring habitats. Species in habitats enclosed by urban sprawl may be similar to those which originally inhabited the site. In newly created habitats, colonising species are determined by the distance they disperse from the nearest source and the conditions prevalent at the site. Many initial colonisers are carried in on the wind, by animals, or in materials deposited on-site. With little competition, primary colonisers gain a foothold, to be replaced eventually by more competitive species. Communities occurring in the early stages are often more species-rich and less common than those found later. The average distance that a species disperses may not be the major factor in determining whether it is able to colonise a site. The occasional extreme distances moved by one or more breeding pairs (or pregnant or parthenogenetic females) are more important, assuming that suitable conditions are encountered. One example of the effect of dispersal ability on colonisation is in the ground beetles (Carabidae). Here, species can be brachypterous (with small wings and

incapable of flight); macropterous (with large wings and able to fly); or dimorphic (with individuals of either form). Macropterous species might be expected to be better able to colonise remote sites. In support of this prediction, a greater proportion of macropterous ground beetles occurred in an isolated urban wasteland (Weigmann, 1982) and more brachypterous species were found at city edges than nearer the centre (Wahlbrink and Zucchi, 1994).

Humans are responsible for the dispersal of many species. The ship rat has spread throughout the world via marine traffic, whilst the rabbit has been deliberately introduced into many areas of the world for food. Many species which are highly successful in Britain are not native: plants such as the sycamore, horse chestnut and more recently rhododendron, and animals including fallow deer and mink have all successfully established following either deliberate or accidental introductions. Invasive species may reduce the conservation value of an area (Usher, 1986) leading to decreased biodiversity, alterations to the geomorphology, hydrology and soil characteristics and possibly directly or indirectly causing the extinction of native species (Lever, 1994). There is evidence that transport networks such as railways and canal systems have helped some species to extend their ranges. Similarly, the transfer of soil and plants around the country enables species to establish in new areas. Soil often contains seeds accumulated from species (especially annuals) where only a proportion of the seeds produced germinate in any one year. This forms a seed bank where a proportion germinate in subsequent years, enabling species to flourish when conditions are appropriate, for example when gaps open in existing plant cover, or when soil is transported to new sites. Although less frequent in animals, a similar example is the eggs of nematode worms.

Arrival is only one aspect of successful col-onisation, although many species may find their way into an area, only a minority survive. Establishment, whether from dispersing populations, introductions or locally present populations (including seed banks), depends on the characteristics of the species concerned. Two extreme strategies are recognised: *r*-selected populations typically have high reproductive rates and produce many eggs, seeds, etc.; and *K*-selected populations which tend to delay reproduction and produce fewer offspring over longer time periods. *r*-selected populations are good colonisers, rapidly establishing in appropriate gaps within habitats (e.g. grasslands). However, they tend to live in habitats which are short lived and may quickly be replaced by the more competitive, if slower growing, *K*-selected populations which are generally in more stable environments (e.g. mature woodlands). Other classifications with three elements have been proposed. In one such (e.g. Grime *et al.*, 1988), plants are classified according to the ability to cope with high disturbance when resources are abundant (ruderal), or low disturbance when resources are plentiful (competitor), or low disturbance when resources are scarce (stress-tolerator). In another system, the relative importance of fecundity, growth and survival on the overall reproductive rate are used to classify species (see Begon *et al.*, 1996 for further details).

Adaptations and Characteristics

Urban greenspaces are harsh environments characterised by sparseness, patchiness, the absence of normally occurring layers of vegetation, and the presence of non-native species (Goldstein-Golding, 1991). Species that can tolerate these conditions often have special adaptations. Tawny owls in urban Poland feed on birds to a greater extent than in rural areas with house sparrows comprising a mere 3 per cent of the diet in rural sites and 88 per cent in the city (Goszczyńsky *et al.*, 1993). This behaviour exploits the availability of small birds in habitats where mammals, the preferred rural diet, are less common. In extreme cases, where the area is dominated by hard surfaces and lacks vegetation, native animals are replaced by synanthropic species (those closely associated with humans) including feral pigeons, starlings, rats and mice. Birds become less diverse and are dominated by omnivores and those adapted to nesting and roosting in artificial structures. Figure 1.1 shows the separation between birds which survive in highly urban areas and those more common in rural habitats. In a study of 848 species of birds from the United States and eastern Asia, McClure (1989) identified characteristics held in common by at least half of the seventy species he categorised as urban. The twenty-nine factors identified included: non-colonial nesting; both parents being involved in nest building and/or rearing young; non-migratory and occurring in groups during at least part of the year; feeding individually or in small groups, on the ground or in low vegetation; tendency to be active and obvious, and being permanent residents. The species best covered by these criteria is the house sparrow (Species Box 1.1) which could be described as the most ubiquitous urban bird.

In central city sites there is a tendency towards smaller species, for example in birds (Cousins, 1982) and arthropods such as ground beetles (Wahlbrink and Zucchi, 1994). For birds this is possibly due to central greensites being small and therefore lacking in resources (although human food waste attracts larger species such as gulls and pigeons). As well as being small, arthropods in

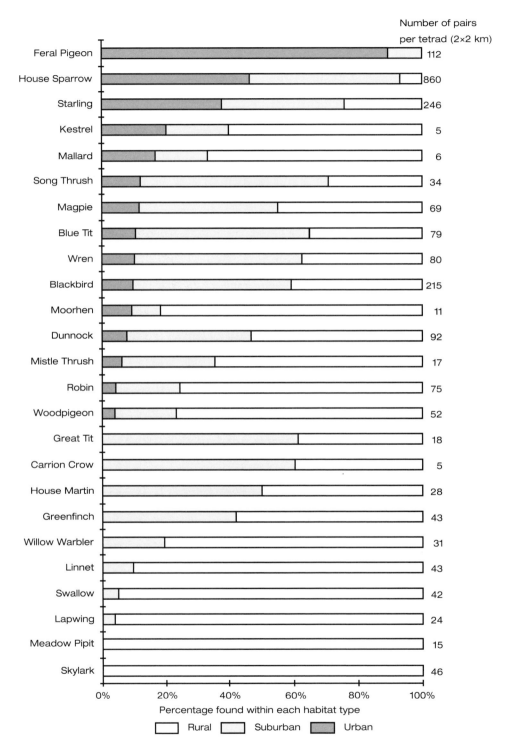

Figure 1.1: Densities of common birds in Greater Manchester (1979–1981). Different species differ in the extent to which they are found in urban and suburban habitats. Data from Holland *et al.* (1984).

··

Species Box 1.1: House Sparrow

The house sparrow is a very common urban and suburban bird around 145 mm in length. Males have a dark grey crown, chestnut nape, pale grey cheeks and a black bib to neck and upper chest, while females and juveniles are dull brown with whitish breasts. House sparrows are widely distributed, especially in human settlements. They nest in both natural (in trees) and artificial (under the eaves of buildings) sites, and there is some evidence that holes are preferred to building nests in the open. Breeding takes place from March to September with two or three broods per year. Commonly four eggs are laid which are incubated by both sexes and take around two weeks to hatch. The young are fed on insects by the parents and usually leave the nest after about two weeks. Adults feed on seeds supplemented by food scraps accidentally or deliberately supplied by people.

Sources: Cramp (1994); Summers-Smith (1988)

··

urban parks and around street trees in Germany tend to be tolerant of a wide range of habitats, highly vagile, with good dispersal capabilities (Schaefer, 1982). The lack of leaf litter, dead wood, stones on the ground, and high disturbance including trampling and pollution are major factors influencing the distribution of urban arthropods.

One example of changing characteristics of urban species is that of industrial melanism. This is where species which occur in more than one form (morph), comprising different colours, tend to be more frequent in the darker form in urban areas compared to rural sites. In one such species, the peppered moth (Species Box 1.2), the population was originally mainly comprised of a light form. From the mid-1800s to 1890 the darker form began to dominate urban populations, especially in heavily industrialised areas including the north of England and the Midlands. The ratios of the

dark: light forms increased from rural to urban sites. Pollution (i.e. smoke from fossil fuels) did not seem to directly influence the moth. It was suggested that the paler form had been camouflaged against lichen on tree trunks. Once a combination of·high levels of sulphur dioxide killed the lichens and soot deposition blackened the bark, the darker morph was then better camouflaged from predators using vision, such as birds. Although this explanation was accepted for some time, doubt has now been placed on it as a complete answer. As urban fuels became smokeless and levels of sulphur dioxide and soot fell during the 1970s, the darker morph has been replaced by the paler one. However, the lichen cover has not changed appreciatively and, in any case, peppered moths rarely rest on tree trunks. For a thorough examination of the case of the peppered moth see Majerus (1989). In other melanic insects, camouflage is not the answer

··

Species Box 1.2: Peppered moth

Peppered moth females are larger (wingspan 53–61 mm) than males (wingspan 41–54 mm). These moths occur in different colour morphs. The typical form is pale with scattered black dots, lines and patches especially concentrated along the leading edges of the forewings. Darker forms (the very dark form *carbonaria* and less dark form *insularia*) have been associated with areas of industrial pollution (see text for discussion). Adults occur in woodlands, parks and gardens from mid-May until mid-August. Eggs are laid in June and the green (or greyish-brown) larvae emerge in July to feed on a variety of trees and shrubs, especially ash, birch, lime and oak, and sometimes on fruit trees. Larvae are cryptic (camouflaged against the background), and resemble twigs. Both adults and larvae are nocturnal. Larvae pupate in September or October in or on the surface of the ground, covered by soil. The peppered moth overwinters as a pupa, with the adult emerging the following spring.

Sources: Emmet and Heath (1991); Skinner (1984)

··

since distasteful animals such as ladybirds (Coccinellidae), which use their bright colours to advertise their distastefulness, may also show industrial melanism. Since darker colours absorb heat better than paler ones, melanic forms may be better able to absorb solar radiation in areas of low sunshine (Stewart and Dixon, 1989). In support of this hypothesis, melanism appears to be restricted to small species such as two-spot ladybirds, rather than larger ones, such as seven-spot ladybirds. Smaller objects lose heat faster than larger ones, and large, dark animals risk overheating on sunny days.

GEOGRAPHICAL INFLUENCES

Some habitat patches behave like islands surrounded by a sea of unsuitable or less suitable habitats. MacArthur and Wilson's (1967) equilibrium theory of island biogeography states that larger islands (whether oceanic or habitat islands) recruit more species and have lower extinction rates than smaller islands. In addition, islands far from sources of potential colonisers or recolonisers (in relation to the dispersal capabilities of the species concerned) have fewer species. This has implications for patchy and fragmented habitats in urban areas. Increasing fragmentation caused by developments isolates populations still further and reduces the likelihood of movement between patches. Larger patches may have larger populations which are better protected against localised extinctions, and larger species are more likely to be in large areas. Greenspaces within large urban areas may behave like remote islands with little connectivity and high extinction and low immigration rates.

Older conurbations tend to be larger and more remote from surrounding rural environments, so reducing colonisation. An isolated wasteland in Berlin had fewer arthropods, especially those with limited dispersal capabilities, than did areas closer to potential sources of colonisers (Weigman, 1982). Size and isolation of urban woodlands are major factors affecting bird diversity (Tilghman, 1987). However, for some birds there is a minimum area below which nesting does not take place and other factors, including building density, are also important for individual species. The species/area effect may be partly responsible for the fact that inner-city areas often have fewer species than suburban ones. In London, there is a trend of decreasing species richness of ground-living arthropods from the edge to the centre (Davis, 1982a). This may also be due to increasing disturbance (including pollution), lower habitat diversity, or other urban factors (including the heat island effect).

Small areas should not be dismissed as lacking wildlife value, especially if they add to the overall habitat diversity of the area. They can even override the species/area effect if several small sites contain more species than one large site of equivalent area. In studies in Oxford on several vertebrate groups, fewer species were found in individual large habitat patches than were recorded in total from groups of several small patches of equivalent combined area (birds – Ford, 1987; amphibians, reptiles and mammals – Dickman, 1987).

The more fragmented habitats are, the greater the proportion of edge to centre. Edges may have different conditions to the centre. For example, woodland edges are less sheltered, sunnier and often drier than non-edge habitats. Since edge effects of between 12–60 m exist for woodland patches, urban woods need to be 450–11,000 m^2 in area to include even small amounts of interior woodland conditions as well as edge habitats (Goldstein-Golding, 1991). Ecotones (transition zones between two or more distinct communities) can be important habitats in their own right. For example, woodland and grassland are often separated by scrub communities which provide conditions for a wider range of plants and animals than are in either of the communities they separate. Similarly, hydrosere communities (the sequence of communities from truly aquatic to terrestrial environments) are very diverse. A wide ecotone may help to reduce human impact and act as a buffer zone for breeding animals. However, transition zones in urban areas are often quite narrow, for example managed areas around ponds, lakes and reservoirs; road verges and railway embankments; and woodland edges in parks. Because some birds and mammals avoid sharply defined edges, abrupt changes in habitat may be a barrier to dispersal (Yahner, 1988).

Many populations of organisms are patchy, linked by occasional dispersal of individuals typically across habitats where conditions are not suitable for them to feed or breed. These are called metapopulations. For example, some small mammals exploit woodland patches which are not directly linked to neighbouring woods. At any one time, not all potentially suitable areas will be occupied by a subpopulation. Localised extinctions occur due to changes in food supply, climatic conditions, or disease. In time, and in the absence of habitat change, small mammals may recolonise from other occupied patches, although frequent extinctions are only rarely replaced. If the unoccupied habitat changes in some profound way (as far as the small mammal is concerned), then even if individuals reach the unoccupied patch, establishment may not be successful.

The degree to which sites are connected via a network of 'stepping-stones' or corridors is important for dispersal. Linear habitats such as road verges, railways and the banks of rivers and canals act as habitats in their own right and as thoroughfares for some species (Spellerberg and Gaywood, 1993). However, the management of corridors is problematic. The large perimeter to area ratio of linear habitats leads to domination by external influences (such as increased insolation, impact of wind, spray drift of agrochemicals and associated increased mortality). Several studies indicate that the width of a habitat corridor may determine the number of species present (see Yahner, 1988). Other determinants are the dispersal capabilities of the species concerned: short distances between open-water habitats (acting as stepping-stones) are less problematic for animals like waterbirds than for those with weaker powers of dispersal such as frogs, which may require corridors.

2

THE ECOLOGY OF URBAN HABITATS

•

Despite the pressures that human populations place on any land found within towns and cities, there are substantial areas of land which have some value to nature, and even some sites with major nature conservation value. The variety of these sites, both terrestrial and aquatic, can be quite amazing; from small derelict sites to large historical deer parks, and from fragments of long-standing marsh to extensive reservoirs. This chapter covers the important features of many of the types of sites found in urban areas.

TERRESTRIAL URBAN HABITATS

The range of habitats in urban areas is surprising. Indeed, most of the major terrestrial habitat types are represented in towns and cities either as remnants of previously rural environments, or as artificial analogues of semi-natural habitats. Details of the most important terrestrial habitats are discussed in this section.

Woods, Trees and Parklands

The built environment is enhanced by trees which are valuable in landscape and amenity, and for screening, aesthetics and wildlife. Woodland in towns and cities is often limited in size and frequently derives from scrub development on land encapsulated but not managed, or land once developed and now derelict. The tree species in towns depend on the history, management and location. In many cases, native species have given way to exotic species, especially along streets and in urban parks.

Several initiatives are planned to increase woodland cover in England, including the creation of a new National Forest in the Midlands and twelve community forests near urban areas. The former will link several towns including Leicester and Burton-upon-Trent and will be accessible to many people. The community forests will link other urban areas (Figure 2.1), for example, the Mersey Forest area includes Merseyside (see case study) and parts of Cheshire, and Red Rose Forest links and extends existing forested sites in the centre, west and north of Greater Manchester. Similar initiatives are planned for Scotland and Wales.

Woodland

Urban woodlands are semi-natural at best, and often the result of development encapsulating remnant wooded sites. Mature woodlands have a continuity of use, leading to stable communities and undisturbed soils. Oak woodland (or beech in the south) is the typical natural type in much of Britain. Within these woods ash, hornbeam, birch and alder may occur together with understorey trees of rowan and holly. Dense shrub layers in these

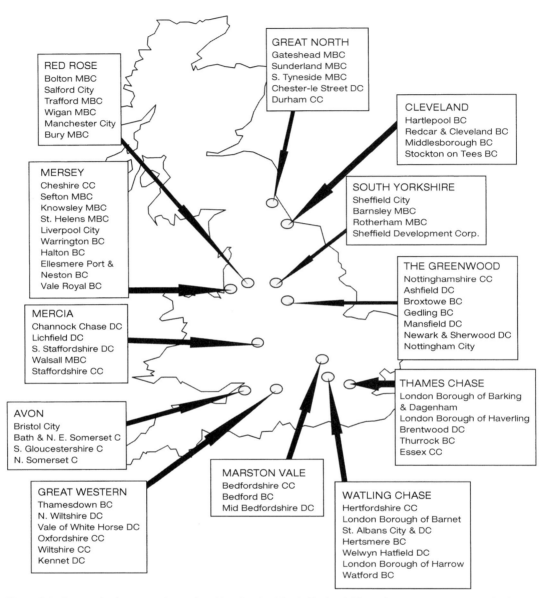

RED ROSE
Bolton MBC
Salford City
Trafford MBC
Wigan MBC
Manchester City
Bury MBC

GREAT NORTH
Gateshead MBC
Sunderland MBC
S. Tyneside MBC
Chester-le Street DC
Durham CC

CLEVELAND
Hartlepool BC
Redcar & Cleveland BC
Middlesborough BC
Stockton on Tees BC

MERSEY
Cheshire CC
Sefton MBC
Knowsley MBC
St. Helens MBC
Liverpool City
Warrington BC
Halton BC
Ellesmere Port &
Neston BC
Vale Royal BC

SOUTH YORKSHIRE
Sheffield City
Barnsley MBC
Rotherham MBC
Sheffield Development Corp.

THE GREENWOOD
Nottinghamshire CC
Ashfield DC
Broxtowe BC
Gedling BC
Mansfield DC
Newark & Sherwood DC
Nottingham City

MERCIA
Channock Chase DC
Lichfield DC
S. Staffordshire DC
Walsall MBC
Staffordshire CC

THAMES CHASE
London Borough of Barking
& Dagenham
London Borough of Haverling
Brentwood DC
Thurrock BC
Essex CC

AVON
Bristol City
Bath & N. E. Somerset C
S. Gloucestershire C
N. Somerset C

MARSTON VALE
Bedfordshire CC
Bedford BC
Mid Bedfordshire DC

GREAT WESTERN
Thamesdown BC
N. Wiltshire DC
Vale of White Horse DC
Oxfordshire CC
Wiltshire CC
Kennet DC

WATLING CHASE
Hertfordshire CC
London Borough of Barnet
St. Albans City & DC
Hertsmere BC
Welwyn Hatfield DC
London Borough of Harrow
Watford BC

Figure 2.1: Community forests and associated local authorities in England. Most of the proposed community forests will be situated near to urban areas. Information from the Countryside Commission.

woods often contain hazel (especially in the south) with some hawthorn, elder and black-thorn. Where dense communities of bracken and bramble have not developed, the ground vegetation may be sparse consisting of mosses, liverworts and lichens. Ground flora in gaps caused by cutting or tree fall is dominated by spring flowers such as bluebells, wood ane-mones and sweet violets, followed by later flowering plants including wood-sorrel, dog's mercury and foxgloves. Ash is a common secondary component of many woodlands.

Wetter woodlands are often dominated by alder and willow.

Woodland and scrub in urban areas are important for small mammals. A survey of fifty patches of woodland and scrub in Oxford found four species of small mammal to be common inhabitants (Dickman and Doncaster, 1987). Wood mice were the most widely distributed, being found in 84 per cent of patches, missing only from parks and allotments. Bank voles were found in 56 per cent of patches and in all types except allotments. Field voles and common shrews were found in about 40 per cent of patches, mainly within woodland, scrub and long grass. Some species were at higher densities than in rural areas (81 ha^{-1} in Oxford and up to 65 ha^{-1} elsewhere for wood mice, and 66 ha^{-1} in Oxford and 10 ha^{-1} elsewhere for common shrews). These high densities are due to confined populations being inhibited from dispersing (thereby increasing density), and reduced predation. Cats are infrequent and foxes in the area take only about 5 per cent (by weight of diet) of small mammals. Rubble and rubbish also provide shelter from inclement weather and predators. Studies in Sheffield (Clinging and Whiteley, 1980) and Manchester (Yalden, 1980a) found wood mice to be the most ubiquitous species. In some woods in Sheffield, bank voles were very common, while in Manchester they were absent even from those habitats (e.g. railway embankments) where they might be expected, possibly reflecting their low dispersal rates compared to wood mice.

Some urban fragments of semi-natural woodland provide suitable habitats for badgers. However, badgers also exploit gardens, sports grounds, and even industrial sites. Urban badger populations are particularly large in Essex, London, Bristol and Bath (Harris, 1984). However, badger populations in London and Bristol have been extensively studied for many years; other cities are less well studied. Recent surveys around Manchester, previously thought to have few urban badgers, identify many setts, mainly on the urban fringe, but with some associated gardens and industrial areas (Wright, 1997). Harris and Cresswell (1987) found a high badger density (4.4 to 7.5 adults per km^2) in suburban Bristol, which is higher than in some rural populations (Cresswell *et al.*, 1990). Urban badgers differ in their behaviour and ecology to rural populations; the population dynamics of the Bristol population are different from those of a high-density rural population in Gloucestershire. The Bristol population had a higher adult mortality, especially in females, and a later female breeding age. There are differences in territoriality between urban and rural populations, with urban badgers often having less well-defined territory boundaries. Urban badgers may be under greater pressure from diggers, especially in the north of England. In Manchester the proportion of disused setts, some which were abandoned following wilful persecution, is higher in urban areas than urban fringe or rural sites.

Many birds, including rooks, jays, woodpigeons and tawny owls, breed in mature trees. The seeds and fruits of trees, shrubs and herbaceous plants attract thrushes and finches, while larger trees house woodpeckers. Trees, whether dead or alive, are very important for a wide range of invertebrates. Sap feeders (such as aphids and other bugs) and leaf-mining moth larvae feed on the leaves, sap runs (leaks from splits in the bark) support a range of beetles and flies, and pollen and nectar from flowers is collected by bees and hoverflies. Ants, spiders and harvestmen hunt for prey. Dead and decaying wood provides homes for flies, beetles, moths and some Hymenoptera such as solitary wasps.

Urban Forestry

Urban forestry includes scrub and secondary woodland colonising previously open areas, as well as trees planted in streets, parks, gardens, woodland and land previously used by industry, and excludes encapsulated woodland or parkland. In Britain, tree planting has only recently been enthusiastically embraced. When new towns (e.g. Milton Keynes) were developed in England and Wales, trees were integrated into the urban landscape to a greater extent than in older towns and cities. More recent initiatives aim to extend these developments in the future.

Urban trees provide local benefits for wildlife; birds, mammals and invertebrates use urban trees as roosts, breeding sites and feeding areas. The numbers of insects supported by trees is important since many urban birds are insectivorous. The number of insect species on each tree species depends on the abundance of the trees and whether they are introduced or native (Kennedy and Southwood, 1984). Thus introduced species generally have few invertebrate species (but see discussion of sycamore later). Local climatic conditions are altered by urban trees. Transpiration, shading and wind sheltering modify local temperatures (Colour Plate 1). Deciduous trees, such as London planes (Species Box 2.1), provide a greater cooling effect than do conifers (Mao *et al.*, 1993). Soil erosion and extremes of soil temperature are reduced by vegetation. Trees also act as watersheds, reducing the amount of rain water which needs to be disposed of in drainage systems. Atmospheric pollution is also ameliorated by trees (see Figure 2.2), particularly since particulate material collects on leaves and fissures in the bark and is then washed into the soil. In urban settings, deciduous trees can reduce dustfall by 27 per cent and intercept 9 per cent of suspended particles in the air (Dochinger, 1980). Coniferous trees have even greater capture rates, reducing dustfall by 38 per cent and intercepting 13 per cent of suspended particles. Other pollutants such as nitrogen oxides, sulphur dioxide, carbon monoxide, ozone and halogens are removed by trees, although some (especially sulphur dioxide, fluoride and ozone) damage trees (Dochinger, 1988). Urban trees reduce noise pollution, with dense rows of evergreen species being more effective than deciduous trees (Mao *et al.*, 1993).

Amenity and screening trees are often badly managed and grow poorly. Trees within urban areas are often under greater stress than they are in rural stands, and street trees in particular face several problems, which are summarised in Figure 2.3. Most problems can be alleviated by better management practices. Soil is often lacking around the roots, and that present is usually colonised by weeds which compete for nutrients and moisture. Rainfall is directed away into drainage systems and soils often have weak water-holding capacities. Growth is reduced by both accidental damage and vandalism (Plate 2.1), and toxic effects of salt and oils from roads and atmospheric pollution including the deposition of soot and other particles onto the leaves. Buildings reduce light intensity and tend to increase wind intensity.

It has been suggested that all of these environmental stresses may reduce the capacity of trees to deal with outbreaks of insect pests. However, this may not be a big problem since trees with little environmental stress in urban parks or ornamental situations receive similar damage from insect herbivores as do equivalent species in natural settings (Nuckols and Connor, 1995). Indeed, chewing insects damage foliage more in natural than in urban woodlands. The major chewing insects (mainly moth larvae) are less likely than sap suckers or gall formers to be introduced into

··

Species Box 2.1: London plane trees

London plane trees are thought to have originated as an introduced hybrid between Oriental
and American plane trees (in around 1675)
planted as street and park trees especially in the
south. Over half the street trees in London are
plane trees, and its development from saplings,
especially in the outer suburbs, is rapidly establishing it as a naturalised London tree. The fact
that many young trees germinated from fertile
seeds has led to some doubt about its hybrid
status (since hybrids are not normally fertile),
and it may simply be a variant of the Oriental
plane tree. London plane trees grow to 44 m tall,
although they are often smaller in street situations. They have large, sharply pointed, three-
to seven-lobed, light green leaves which persist
in leaf litter throughout winter. The olive-grey
bark peels in large patches on the trunk and
major branches to reveal paler, creamy bark
beneath. The leathery leaves are easily washed
clean of grime by rainfall, and this, together with
its shedding of soot covered bark, late leaf formation and early leaf fall, helps the tree survive
smoky and polluted environments. Plane trees
cope well with restrictions on root space, as
occurs on streets. Flowering occurs in May, and
the male and female flowers look similar and
grow on the same tree but in separate clusters.
The wind-carried fruit consist of four seeds,
each with a tuft of yellow hairs.

Sources: Stace (1997); Edlin and Mitchell (1985)

··

urban trees, due to low dispersal into urban
areas and a lower chance of being introduced
in soils or planted trees. In Hong Kong, most
street trees suffer from structural or physiological problems, with large specimens having
more problems (Jim, 1992). Pest damage was
rarely serious, except in weakened individuals.
Many trees were growing under extremely
cramped conditions, where planting of rapid
growing, tall but narrow-canopied species
instead would be advantageous.

The frequent failure of urban tree planting is
often due to a lack of attention to conditions
on the ground, insufficient monitoring and the
use of inappropriate species (Hodge, 1993).
Protecting against damage caused by the public and rabbits, is important for successful
establishment. Harsher environments require
more tolerant species, and even where conditions are mild, planting mixtures which
include some very tolerant ones will ensure
some tree cover. Increasingly, species are
selected for appropriate characteristics (as
shelter belt trees, ornamentals, etc.) and their

Atmospheric pollutants such as sulphur dioxide, chlorine and fluorides may enter the stomata (pores on the leaf surface) where they are used or released into the air or soil

Pollutants may be trapped by spines, hairs, waxy surfaces and films of moisture on the leaves

Winds passing through the canopy lose speed and some of the pollutants they are carrying

Evaporation from the leaves cools the surrounding area

Evergreen trees may have a larger surface area of leaf and keep their leaves for longer than deciduous trees

The number, pattern and density of the leaves influence the potential to reduce air pollution

A canopy with many branches and a large trunk with fissured bark will trap more particles from the air

Soil around trees acts as a major depositing area for pollutants including heavy metals and carbon monoxide

Figure 2.2: The air-cleansing action of trees.
Sources: Dochinger (1980, 1988)

tolerance to urban conditions (e.g. pollution, dry soils). Although it is preferable to use native species wherever possible (see Chapter 3), non-native species and varieties possess characteristics making them more suitable for some uses as described in Table 2.1. In general the larger the woodland, the more native species tend to be used.

Wooded Grounds and Parkland

Parkland, as opposed to town parks, is often associated with a large house, and frequently originates from early deer parks. Deer parks in Britain began in Norman times and increased during the twelfth century, probably with the introduction of fallow deer (Rackham, 1986). Today, the commonest deer in British parks are the indigenous red and roe deer, and the introduced fallow, sika, Chinese water deer and Chinese muntjac. The last three of these are relative newcomers, having been introduced at the end of the nineteenth and beginning of the twentieth centuries. Fallow deer (Species Box 2.2) were introduced by the Normans and are the traditional deer kept in

Plate 2.1: Vandalised street tree

parks. Their grazing habit suits them to parkland to a greater extent than browsers such as red deer. Some mediaeval parks became landscape parks during the eighteenth century when the grounds of country houses were enhanced by the park designers of the day, often utilising some older trees in the landscaping process, a feature replicated in more recent parks. Much of the parkland within urban areas is encapsulated at the fringe. Although much parkland within or adjoining urban areas no longer houses deer (e.g. Hyde Park, St James's Park; and Regent's Park in London), some still does (including Lyme Park near Stockport, Ashton Court near Bristol and Richmond Park in London).

Parks, Gardens and Allotments

Parks and gardens can comprise a large proportion of a town or city, especially in the suburbs. For example, of the 27,000 ha covered by the City of Birmingham, around a ninth is parks and open spaces, and there are an estimated 250,000 domestic gardens (Land Care Associates, 1995). Despite many of the plants being introduced species, open space in otherwise built-up areas offers shelter and food for invertebrates, birds and some mammals. Mixtures of species which between them produce flowers and fruit throughout the year encourage a surprising range of species. Formal plantings near the centre of Manchester include ornamental species (such as Oregon-grapes) which produce nectar over winter. Honeybees from local hives feed on sunny January days, gaining enough nutrients that they rarely need to be given extra food. This winter nectar, together with that produced by other species in autumn (balsam and willow-herb) and spring (cotoneaster), provides food all year round for many flying insects such as bumblebees and hoverflies. Birds also benefit from exotic shrubs; in winter waxwings feed on berries in gardens and, in one area of Manchester, have been seen in bushes alongside busy urban streets.

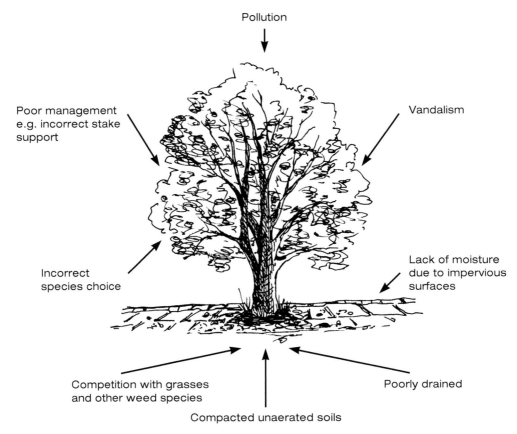

Pollution

Poor management
e.g. incorrect stake
support

Vandalism

Incorrect
species choice

Lack of moisture
due to impervious
surfaces

Competition with grasses
and other weed species

Poorly drained

Compacted unaerated soils

Figure 2.3: Reasons for the failure of street trees.
Source: Hibberd (1989)

Parks

The town-park movement in the nineteenth and twentieth centuries provided ornamental parks for urban dwellers. Initially, many were created by local philanthropists as pockets of countryside-like open space in which urban dwellers could walk. As the numbers using the parks grew, policies were introduced to keep the public off the grass and on tougher paths. In the latter quarter of the nineteenth century, interest in gardening led to the introduction of formal flower beds and more regimented designs. The timing of the creation of urban parks during, rather than at the beginning of, urban expansion means that many inner-city areas are devoid of these facilities. During the first thirty years of the twentieth century, local authorities attempted to redress this imbalance. The subsequent expansion of parks has continued to a greater or lesser extent throughout the rest of the twentieth century. For example, twenty-nine parks were acquired in Edinburgh from 1945 to 1969, and twenty-seven in Leicester from 1950 to 1975 (Walker and Duffield, 1983). Probably as a result of their origins, many public parks are recreational and aesthetic retreats rather than rich wildlife reserves (Plate 2.2). However, there is an increasing tendency (often stimulated by high costs of existing management) to reduce

Table 2.1: Tree species suitable for urban planting

	Scale of planting		
	Small – less than 0.25 ha or one to 500 trees	Medium – approximately 0.25 ha or about 500 trees	Large – over 0.25 ha or over 500 trees
Types of site	City-centre developments and office blocks Shopping areas Housing estates Gardens and recreation areas Car parks and roundabouts	Parks, golf courses and other recreation areas Road and street developments Housing, industrial and commercial developments Reclaimed industrial land and inner-city developments Railway land	Urban fringe Reclamation sites Abandoned quarries and allotments Major road developments Existing urban woodlands
Possible uses	Decorative, ornamental, visual screening/softening	Landscape improvement, screening and urban greenspace	Timber production, recreation and wildlife conservation
Examples of suitable species	*Acer palmatum* (Sw Sc Sl; Hs; Ad) *Alnus incana* 'Pendula' (Sw Sc Sl Si; Hs Hp) *Betula pendula* 'Youngii' (Sw Sc Ss Si; Hc Hs Hr Hp; Ad) *Corylus avellana* 'Contorta' (Sc Ss Sl; He) *Crataegus laevigata* 'Paul's Scarlet' (Sw Sc Sl; Hc He Hr Hp; Ad) *Fagus sylvatica* 'Aurea Pendula' (Ss Sl Si; Ad) *Prunus* species e.g. 'Kiku-shidare' (Sc Sl; Hc Hp; Ad) *Juniperus virginiana* 'Skyrocket' (Sc Ss Sl; Hs) *Picea abies* 'Inversa' (Sw Sc Ss; He Hs) *Salix caprea* 'Kilmarnock' (Sw Sc Ss Sl Sli; He Hs Hp) *Salix daphnoides* (Sw Sc Sl Si; Hc He Hp Ad) *Sorbus aria* 'Pendula' (Sw Sc Ss Sl Sli; Hc He Hp; Ad) *Sorbus aucuparia* 'Beissneri' (Sw Sc Ss Sl Si; Hc He Hr Hp; Ad)	*Acer platanoides* (Sw Sc Ss Si; Hc He Hr Hp; ad) *Alnus glutinosa* 'Aurea' (Sw Sc Sl Si; Hc He Hs Hp) *Betula pendula* 'Dalecarlica' (Sw Sc Ss Si; Hc He Hs Hr Hp; Ad) *Fagus sylvatica* 'Dawyck' (Ss Sl Si; Hc He Hs Hr; Ad) *Fraxinus excelsior* 'Pendula' (Sw Sc Sl Si; Hc He Hp) *Juniperus communis* (Ss Sl Si; Hc He He) *Prunus avium* 'Plena' (Sc Ss Sl Si; Hc Hr Hp; Ad) *Prunus laurocerasus* (Sc Si; Ah Ae) *Prunus lusitanica* (Sc Si; Ah Ae) *Quercus rubra* (Sw Sc Ss Si; Hp; Ah Ad) *Salix alba vitellina* (Sw Sc Sl; Hc He Hp; Ad) *Sorbus aria* (Sc Ss Sl Si; Hc He Hp; Ad)	*Acer campestre* (Sw Sc Ss Si; Hc He Hp; Ah Ad) *Alnus glutinosa* (Sw Sc Sl Si; Hc He Hs Hp) *Betula pendula* (Sc Ss Si; Hc He Hs Hr Hp; Ad) *Corylus avellana* (Sc Ss; He) *Crataegus monogyna* (Sw Sc Ss Sl Si; Hc He Hp; Ah) *Fagus sylvatica* (Ss Sl Si; Hc He Hp; Ad) *Fraxinus excelsior* (Sc Sl Si; Hc He Hp) *Pinus sylvestris* (Sc Ss Sl Si; He) *Populus nigra* (Sw Sc Sl Si; Hc He Hp) *Populus tremula* (Sw Sc Ss Si; He) *Prunus avium* (Sc Ss Sl Si; Hc Hr Hp; Ad) *Quercus petraea* (Sc Ss Sl; Hc; Ad) *Quercus robur* (Sw Hc Ss; Hc Hp; Ad) *Salix alba* (Sw Sc Sl Si; Hc He Hr Hp; Ad) *Sorbus aucuparia* (Sw Sc Ss Sl Si; Hc He Hr Hp; Ad)

Sources: based on recommendations by Hibberd (1989), with additional information from Bradshaw and Chadwick (1980), Emery (1986), Grime *et al.* (1988) and Stace (1997)

The codes in parentheses following each species name indicate recommendations for particular soils, sites or functions; those aspects for which the species is strongly recommended are double underlined:

Soils type (S)
w – wet soil
c – clay soil
s – dry sandy soil
l – chalk soil
i – industrial spoil

Habitat type (H)
c – coastal sites
e – exposed sites
s – small sites
r – sides of roads or streets
p – sites affected by smoke and fumes

Attributes (A)
h – hedges, screening, shelter belts
d – decorative
e – evergreen broadleaves

* – native species or varieties of native species

Species Box 2.2: Fallow deer

Fallow deer in the summer are reddish-fawn, with white spots along their sides and back and a black stripe extending from the nape to the end of the tail. In winter they are greyish-fawn with less distinct spots. Males are up to 1 m tall at the shoulder and 50–90 kg, while females are smaller (up to 0.9 m and 35–56 kg). The sexes usually stay apart for much of the year, coming together in autumn to breed. However, in open habitats mixed herds may persist throughout the year. The antlers are grown in August/ September and cast in April to June. For about three weeks in September and October males compete for rutting stands (areas where they call to attract females). However, in herds where the sexes mix all year round, males may not establish rutting stands, instead mating with any receptive female encountered. Single young are born in June with lactation lasting up to nine months. Males mature at seven to fourteen months (although they may not breed until later) while females breed annually from sixteen months. In many managed herds, the sex ratio is

artificially kept at about one male to three females. Fallow deer tend to lie up during the day and feed at dusk. They eat grasses, but also trees and shrubs (especially through the winter) and fruits in autumn.

Source: Corbet and Harris (1991)

---

the formalised appearance and return to areas more reminiscent of the countryside.

Although, the majority of plants in parks are cultivated (Colour Plate 2), the animals are not, and may be quite abundant. Species such as grey squirrels (Species Box 2.3) have become common throughout many cities, especially in parks in the suburbs (Colour Plate 3) and even in the heart of the city. More animal species can be found in parks nearer the urban fringe. Even in a good urban park, the number of bird species is unlikely to exceed thirty, compared to up to fifty in a good suburban park and around sixty in a city as a whole (Hounsome, 1979). In London, the number of bird species in parks increases with increasing distance from the centre (Figure 2.4). However, there is a clear distinction between urban and suburban parks, possibly

due to the increased disturbance, more formal nature and smaller sizes of urban parks. In some cases changes in management practice have influenced the species present: surveys in early November of Kensington Gardens, found more species (forty-four) in 1975 compared to 1925 (twenty-six). During this period, park management was tailored towards less formality and greater habitat diversity. For invertebrates such as ground beetles, there are also differences between city and rural parks. In Warsaw, Poland, only a subset of those occurring in rural parks were found in urban parks (Czechowski, 1980).

Gardens and Allotments

Landscaped grounds and gardens have been associated with large houses for centuries,

Plate 2.2: Typical town park

although the expansion of gardens associated with smaller dwellings probably dates from the 1700s. Today, many homes have gardens with lawns and flower beds, which provide food and shelter for wild plants and animals. Private gardens probably comprise around 3 per cent of the land area of England and Wales (Gilbert, 1989), and form fairly continuous tracts in many suburban areas. Hedgehogs (Species Box 2.4) are common in parks and gardens and, although most individual gardens are too small to support them, the area covered by several suburban gardens may be ample. It is possible that those near buildings are less at risk from predation by badgers which tend to avoid residential and more urban areas (Doncaster, 1994). Despite the frequent absence of native plants, gardens usually possess a diversity of vegetation structures (trees, shrubs, herbaceous plants and grasses) together with species producing flowers and fruit over much of the year. When compost heaps, hedges, walls and ponds are considered, gardens have considerable diversity of habitats. Compost heaps are especially

rich in soil invertebrates and well-managed heaps provide warm conditions throughout the year. Allotments are generally more intensively managed than most gardens. However, they still provide a range of habitats for some plants (especially around the edges) and many animals, particularly invertebrates.

Although some invertebrates are considered to be pests in cultivated areas (e.g. slugs, snails and caterpillars of moths and butterflies), many do little harm (e.g. millipedes), are benign (e.g. woodlice), or are beneficial. Amongst the beneficial invertebrates are bees and hoverflies which pollinate plants, earthworms which are involved in soil formation and ground beetles, ladybirds and spiders which predate a range of pest species. Predators include active hunters such as wolf spiders, harvestmen and rove beetles, as well as more passive species such as orb-web-building spiders like the garden spider (Species Box 2.5).

There is not much information on the ecology of gardens, although in-depth surveys of invertebrates have been carried out at some

....

Species Box 2.3: Grey squirrel

Grey squirrels are larger than native red squirrels. They are 240–285 mm long with a further
195–240 mm tail, and weigh around 0.5 kg.
Females are slightly heavier than males. Both
sexes have summer coats which are grey above
with chestnut legs and feet and whitish undersides. In winter the coat is thicker and the legs
are grey. Grey squirrels were introduced from
north America several times from at least 1876.
They expanded rapidly from 1930 until 1945,
since when they have spread more slowly. They
now occur in several wooded habitats and
favour deciduous and mixed woodland, coniferous woods with neighbouring deciduous trees,
as well as urban parks, gardens and areas of
dense scrub. Mating occurs in January and two
to four young are born in February. A second
litter may occur in July. The young begin to leave
the nest after seven weeks, although they are not
weaned for another one to three weeks. They are
sexually mature after ten to twelve months but
rarely produce more than one litter in their first
year. Although they do not actually hibernate,
several animals may gather in a drey (domed
nest) in winter to keep warm. Grey squirrels often
forage on the ground particularly during winter
where they feed on seeds, fruit and buds
(depending on the time of year and availability)

as well as food distributed to them by the public.
They cache food in autumn when it is abundant.
Although an endearing visitor to the garden,
they may cause some nuisance in taking food
from bird tables and stripping bark from trees
(especially beech and sycamore).

Sources: Gurnell (1987); Corbet and Harris (1991)

....

sites, for example Kevan (1945) recorded over
150 species of beetle in a normally cultivated
Edinburgh garden over two years. In an extensive study of one suburban garden in Leicester,
Owen (1991) found many wildflowers (166
species or about 11 per cent of the British
flora), vertebrates (three species of amphibians, seven mammals and forty-nine birds)
and invertebrates (Figure 2.5). This garden
was laid out in 1927, and is large (741 m^2
compared to an average garden size of 186
m^2). Although it is tended with wildlife in
mind, it has not been extensively managed as
a wildlife wilderness and contains a lawn,

flower and vegetable beds and trees and
shrubs. Some plants are grown to attract
insects (butterfly-bushes and wildflowers such
as feverfew, hogweed and willowherbs) or
birds (currants, holly, cotoneaster). Even the
smallest and neatest flower bed is home to
subsoil and soil-surface animals, in addition to
those living amongst the plants.

One animal which has become more common in suburban gardens in recent years is the
fox (Species Box 2.6). In urban areas, foxes
occur most frequently in residential areas, particularly where the density of housing is low
and where houses were predominately built

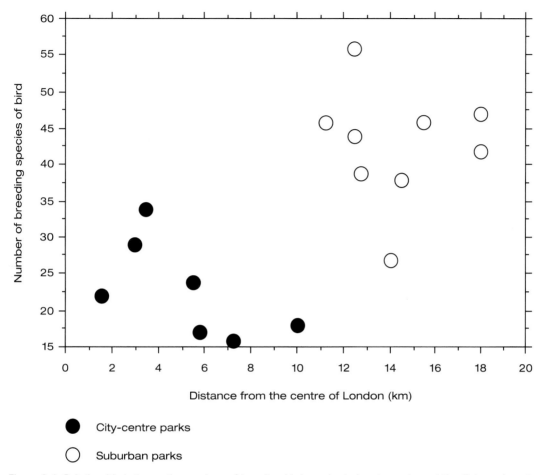

Figure 2.4: Relationship between the numbers of breeding bird species in London parks and the distance from the city centre.
Source: Hounsome (1979)

between the wars (Harris and Rayner, 1986). Some surveys suggest a lower incidence of urban foxes in the north (e.g. Harris and Rayner, 1986), however some northern cities (e.g. Manchester) have large fox populations. Urban foxes sometimes dig earths close to houses and other buildings (Harris, 1981) and in highly disturbed areas including under garden sheds (P. Chipman, unpublished). In rural areas, foxes tend to live in territorial pairs. However, in urban and suburban Oxford adults usually live in groups of three to five often in much smaller territories than their rural counterparts (Doncaster and Macdonald, 1992). The differences in group size between rural and urban areas may be due to the distribution of patches of resources. Areas with sufficient resources for a pair of animals are relatively difficult to defend compared to larger territories which have sufficient resources for bigger groups of adults. Suburban territories tend to be stable, but city living animals change location continually, possibly due to a combination of higher mortality,

Species Box 2.4: Hedgehog

Hedgehogs are the only spiny mammals native to Britain (although porcupines occur in some areas where they have escaped from captivity). At one year old, hedgehogs are about 0.6–0.7 kg in the spring rising to 0.9–1.0 kg in autumn (larger males can reach 1.2 kg). They are about 200–300 mm in length and are squat spiny animals with longish legs partially hidden by a skirt of long hairs. The broad front feet are used for digging. The spines are hollow, modified hairs, 20–30 mm long and 2 mm in diameter, tapering to a point and strengthened by longitudinal ridges. Hedgehogs defend themselves against predation by rolling into a ball with their spines bristling out all over. Young or weak animals may be taken by foxes or badgers, although the biggest threat in urban areas is traffic. Hedgehogs are present throughout Britain, usually close to scrub or woodland. They are mainly nocturnal but can be seen at dawn and dusk. Mating occurs in spring and litters of four to six young are born in June or July. If breeding occurs in late summer it is occasionally due to a second litter, but more often a late first litter. The young are born blind and suckle for four weeks, leaving the nest after about twenty-two days and dispersing after around six weeks. Hedgehogs are insectivores, feeding especially on beetles and caterpillars as well as other invertebrates (including earthworms), birds eggs and occasional small mammals (perhaps as carrion) and birds. Deliberate feeding by people may encourage their presence in gardens. They hibernate in a nest of leaves and grass under scrub, in burrows or under garden sheds from October until the spring. The young hibernate later than the adults, first acquiring sufficient fat reserves to last them through the winter. Consequently, young from late litters have less chance of surviving hibernation.

Sources: Corbet and Harris (1991); Reeve (1994)

disturbance and a more fluctuating food supply. In many cases householders regularly feed urban foxes so reducing their need to forage further. Contrary to popular belief, urban foxes do not often eat pets. Foxes ignore cats (sometimes even feeding together from the same source). Losses of smaller pets such as guinea pigs and rabbits are more often the

Species Box 2.5: Garden spider

The garden spider has a globular, oval (almost shield shaped) body and relatively short legs. Garden spiders are light to dark brown with a row of white spots down the centre of the back and a fainter row at right angles across the back forming a cross set within a darker triangular patch. As in many spider species, females are larger (10–13 mm) than males (4–8 mm). In common with other members of the family they are orb-web spinners and produce a web of up to 400 mm in diameter in vegetation (or sometimes on window frames) at 1.5 to 2 m from the ground. Males spin a courting thread onto the outside of the female's web, and attempt to attract the female by strumming the courting thread with their legs. If the female is receptive, the pair mate. Mating lasts for less than 30 seconds. The male transfers sperm to the female's copulatory organ using the specially adapted pedipalps (foreleg-like appendage normally used for feeding). Females sometimes eat their partners after mating, although it is probably mainly older males that are caught. The eggs are wrapped in a spherical sac of silk and guarded by the female until she dies in autumn. This guarding may reduce the likelihood of attack by parasitoids. Eggs develop in the following spring. Large nursery clusters of orange spiderlings with black triangular markings occur

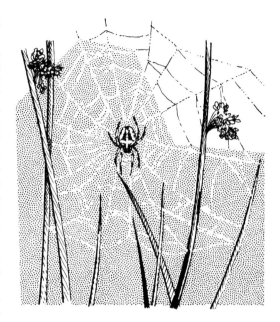

during the summer. These disperse rapidly in all directions if disturbed. The young produce similar webs to the adults, initially catching smaller prey, but quickly consuming animals which are much larger than themselves. Adults are usually found in late summer and autumn.

Sources: Roberts (1985); Foelix (1982)

work of larger dogs, particularly if there is evidence of extensive damage to cages or hutches.

Wood mice are common inhabitants of dense undergrowth in gardens although they may venture inside sheds and greenhouses. House mice usually occur in buildings, but are also found in gardens. Bank voles are more frequent than field voles, except where gardens back onto rough grassland habitats. All four rodents feed on plant material, and some damage vegetables, bulbs and fruit crops. In Sheffield, pygmy shrews are rarer in urban or suburban habitats than are common shrews (Clinging and Whiteley, 1980). Moles, shrews and hedgehogs are insectivores and feed on a range of invertebrates, including many pest species. Moles are useful in that their tunnelling assists the drainage and aeration of soils, although this is unsightly when in the middle of a lawn, and they do eat earthworms which are also useful species.

Gardens provide food, perches and nesting sites for common birds such as blue tits and robins (Species Box 2.7) and rarer urban species such as jays. In one survey there were

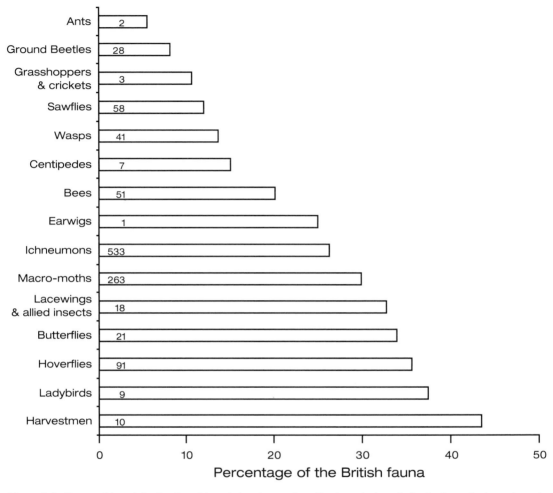

Figure 2.5: Range of invertebrates found in a Leicester garden. Numbers in bars indicate the actual number of species found.

Source: Owen (1991)

more species in rural (20.2 species in winter and 20.5 in summer) compared to suburban gardens (18.2 species in winter and 15.5 in summer) (Glue and Muirhead, 1991). The range of species also differed, although the top six birds in urban gardens (blackbirds, house sparrows, blue tits, starlings, robins and song thrushes) were also amongst the top ten in suburban and rural gardens. Many of these species take supplementary food (all six species being amongst the top twelve feeders in

gardens during winter). One species which has increased in frequency in some areas is the magpie, which in 1994/5 was found in all gardens surveyed in Greater Manchester (Smith, 1996).

Playing Fields, Golf Courses and Other Grasslands

During the 1920s and 1930s, concern for the physical fitness of the British public

Species Box 2.6: Foxes

Foxes are yellow-orange with black fronts to the feet and backs of the ears. Males are on average 1.8 m long (including a tail of 0.41 m) and weigh around 6.7 kg. Females are smaller (1.1 m long; tail 0.39 m) and lighter (5.4 kg). Rural foxes are carnivorous, mainly feeding on small mammals, birds, insects and earthworms. Urban foxes are much more omnivorous, scavenging for over one-third of their food from dustbins. However, rural foxes may also scavenge substantially round villages' and farms. Foxes usually form monogamous pairs in rural but not necessarily in urban situations (see text), and the mating season is during January. Females locate earths in February and the cubs (usually four or five per litter) are born in March, emerging in April. Cubs eat increasing amounts of solid food from May and are able to forage for themselves by August. By the autumn cubs are fully grown and indistinguishable from the adults and the family group begins to disperse. The young are able to reproduce within their first year.

Sources: Macdonald (1987); Corbet and Harris (1991)

culminated in the provision of recreation grounds (Walker and Duffield, 1983). Financial support, together with legislative and educational initiatives, ensured that all schools had appropriate recreational facilities and led to a major expansion in the number of playing fields in towns. Because of their primary purpose, playing fields lack diversity, except for surrounding belts of trees and rough grassland (Plate 2.3). Golf courses, with their water hazards and areas of deep and semi-rough, have a greater scope for diversity. Other grasslands in towns and cities usually exist because of incorporation by urban sprawl, some (such as common land) retained for historical reasons.

Playing Fields

Several species of birds feed on playing fields, including magpies (Species Box 2.8) and gulls,

Species Box 2.7: Robin

Nominated as Britain's national bird, the robin is a small (140 mm in length and 19.5 g) resident of many parks and gardens. Adults have orange faces and breasts with olive-brown heads and backs. Juveniles do not have orange breasts and are mottled dark brown and buff. Robins nest in holes in banks and walls and sometimes in very artificial situations such as household objects, jacket pockets and cars. The female builds a nest of grasses, leaves and moss. Robins breed in late March to the end of July, producing two clutches per year of five to seven eggs, although failed clutches may be replaced, to produce up to three clutches in a year. The female incubates the eggs for around two weeks whilst being fed by the male. The young take around two weeks to fledge, during which time they are fed, first by the male and then by both parents. Robins feed on invertebrates (including insects, spiders and earthworms), seeds, soft fruit and berries. They are mainly ground feeders, using perches from which to locate potential food. In Britain, most males and some females hold winter territories whilst a few (mainly females and juveniles) migrate south for winter.

Sources: Cramp (1988); Lack (1972)

both of which have become much more abundant in recent years. Magpies are particularly successful in urban areas, although numbers are also increasing in rural and suburban habitats (Stroud and Glue, 1991). In Greater Manchester, magpies were found in thirty (out of thirty-two) 1 km squares in 1995 compared to twenty-five squares in 1994 (Smith, 1996). Magpies have been implicated in the reduction of songbirds over the last few years. However, vertebrate prey is not very important in magpie diet, and nestlings and eggs are only occasionally taken. Additionally, there was no evidence for a magpie-related decline in other birds in a seven-year study in north-west Germany (Kooiker, 1994). Indeed, although

magpie numbers more than tripled (from 2.54 to 9.66 nests 100 ha^{-1}), the number of other birds also increased (by 40 per cent, although only by 19 per cent if woodpigeons are excluded). Compared to rural populations, urban magpies lay their eggs earlier, a habit which may be facilitated by their tendency to reuse old nests and the enhanced early food supply due to the mild urban climate (Tatner, 1982).

Several species of gulls successfully exploit urban areas, in some cases (e.g. black-headed gulls) to the extent that the inland population exceeds that at the coast (Stroud and Glue, 1991). Both black-headed and common gulls are frequent on open grassland, although they

Plate 2.3: Closely mown school playing field

use the area differently, with black-headed gulls being more mobile and feeding less on playing fields than do common gulls (Jones, 1985). In both species, juveniles show a greater turnover at feeding sites than do adults. Juveniles use a wider range of feeding sites and go further afield than adults, possibly to visit areas which may have potential as future feeding and breeding sites.

Golf Courses

There are over 2,500 golf courses in Britain, covering about 1,500 km²; at least 0.5 per cent of the land area (Tobin and Taylor, 1996). Large areas are manicured (greens and fairways), but there is a substantial amount of deeper rough (about 35 per cent of the total) which has some importance for wildlife. Trees and hedges are used for nesting, roosting and food sources for several species of birds. Uncut ground vegetation in the rough houses wild flowers, invertebrates and small mammals. Ponds and other water courses can be rich in invertebrate, plant and bird life. Many older,

inland courses developed on parkland and heathland, and thus contain more areas of value to wildlife than those recently built on farmland or reclaimed sites. Some golf courses provide habitats for rare species, for example natterjack toads and sand lizards are present at Royal Birkdale near Liverpool. Other courses support chalk grassland, dune slacks or ancient woodland, together with their associated flora and fauna (including many orchids, butterflies and bats). In 1987, eighty-one courses had a part designated as a SSSI.

Other Grassland Habitats

Some fragments of grassland and heathland were encircled by urban sprawl. These are often fairly small, although they may be quite diverse and important habitats, especially at a local or regional level. Lowland heathlands are mainly in the south of Britain and are typified by acidic, infertile soils supporting communities of heather and other dwarf shrubs, including bell heather and cross-leaved heath. Human activity has fragmented large areas of

Species Box 2.8: Magpie

Magpies are distinctive black and white birds about 450 mm in length with purple/blue iridescent markings on the body and a long green tail. They occur in small groups in wooded areas, open habitats with scattered trees and increasingly in urban areas even in the absence of trees. They defend territories of about 5 ha and tend to have regularly spaced nests (breeding at up to 30 pairs per km^2). Magpies produce bulky nests of sticks and mud with hair and plant material inside, at the tops of tall trees and bushes. Most nests have a domed roof and a single entrance, although up to 30 per cent are open (with inexperienced birds being more likely to build open nests). They have a generalist diet of large insects (usually taken from or just under the ground), carrion, seeds, grain, nuts and berries, supplemented by occasional small mammals, eggs and young birds. In summer they feed mainly on invertebrates, whilst in winter they eat more plant material. They often cache excess food under ground. Breeding occurs from the end of March to June, usually with a single brood of about six eggs which hatch asynchronously following an incubation period of around twenty-one days. The smallest chicks are least likely to survive unless food is plentiful and usually only two or three survive to fledge twenty-seven days later. Most subsequent

mortality (up to 66 per cent) occurs within a few months after fledging, after which survivorship is fairly constant over time with birds living an average of two years (females) to three and a half years (males).

Sources: Cramp (1994); Birkhead (1991)

former heathland, many of which have been lost to urban development. Around London, many former heathlands south-east of the Thames are built on, while to the north-east of the river isolated fragments remain, several of which are now public open spaces (Webb, 1985). One such important area of remnant heathland lies within Wimbledon Common. Commons frequently comprise semi-natural habitats under-represented in the rest of the locality. For example, commons in Herefordshire contain half the chalk grassland and over a quarter of all unimproved and semi-

improved grassland in the county (Penfold and Francis, 1991). Similarly, in West Glamorgan almost all remaining heathland is on common land, and in Radnor (now part of Powys) there is little heather moorland except on upland commons.

Common land was originally privately owned land, often outside a town, which had common grazing and food collection rights for local people. There are 8,674 commons in England and Wales with a total area of over 5,500 km^2 (Aitchison, 1990), comprising about 4 per cent of the total land area (Penfold and Francis, 1991). The north and west have the most commons, although the largest area of common ground is in the south-east. In Scotland and Northern Ireland there are no designated commons, but there is other land (sometimes covering large areas) on which local communities have agricultural rights. Usually commons associated with urban areas are small, but there are exceptions such as the 1,000 ha of Sutton Park in Birmingham. Commons are an important wildlife refuge, with 31 per cent of the land area of commons being designated as SSSIs (Penfold and Francis, 1991). The high conservation value is mainly due to a history of relatively low-intensity agriculture and legal protection. Common rights are now more concerned with public access than grazing. However, often as a result of disuse, several commons, especially in urban areas, have lost most of their rights (78 per cent of English commons have one or no rights), putting them at risk of deregistration and development.

On many commons, grassland and woodland are separated by a soft edge. This contrasts with the sharper ecotones more usual in urban habitats. Rougher grassland areas also include scrub communities of hawthorn, gorse and bracken. This habitat diversity (sometimes including aquatic habitats) provides a wealth of opportunities for wildlife in urban areas, occasionally even in central city sites. Over 400 species of vascular plants, 100 species of birds and twenty-five species of butterflies have been recorded on Southampton Common, and similar diversities have been found on Wimbledon Common and on Clifton Down in Bristol (Laurie, 1979). These large sites have substantial potential for maintaining remnant populations, and acting as sources for the colonisation and recolonisation of other urban greenspace. Part of Wimbledon Common, for example, is a SSSI covering heath-

land, grassland, scrub and woodland habitats. Grassland plants attract many invertebrates: butterflies including the holly blue; other flying insects such as hoverflies; snails (e.g. *Cepeae nemoralis* and *C. hortensis*) and orb-web-spinning spiders (e.g. *Cyclosa conica*). Disturbance is often an important factor influencing these communities. On Hampstead Heath, where disturbance is high, only a few species, but many individuals, of spider use surface habitats (Milner, 1993). Many species in disturbed areas are common; more diverse communities containing rarer species are restricted to areas which are protected and managed less intensively; in the case of Hampstead Heath to protected areas within a bird sanctuary.

Demolition Sites

Gilbert (1989 and 1992a) described another type of urban common: an area which was once a demolition site or similar and is now naturally colonised by plant communities which are definitively urban in character. Demolition of buildings and subsequent levelling produces bare substrates composed of building rubble (usually brick or concrete) mixed with finer material (often containing a large amount of mortar). These can quickly be colonised by plants (Plate 2.4). The conditions are often not uniform across a site, but are usually low in organic material, reasonably fertile (though often lacking nitrogen), rapidly draining and alkaline.

Plant communities on urban commons differ regionally, with species such as the butterfly-bush being frequent in the south and rarer in the north-east of England (Colour Plate 4). The drier east is dominated by false oat-grass, creeping thistles, hogweed, carrot and wild parsnip. In wetter western sites, umbellifers (like hogweed and carrot) are less

Plate 2.4: Plants colonising building rubble

frequent, but Yorkshire-fog and reed canary-grass are abundant, with rushes occurring in the wettest soils. In the west, Japanese knotweed may dominate. In Scotland where conditions are cool and wet, rhododendron, heather, sedges and mosses are common. Species composition is influenced by many other factors. The soil of surrounding districts determines the flora available as colonists. Dumped rubbish, rich in organic materials, encourages species typical of fertile habitats (e.g. common nettles and ground-elder), whilst other species are introduced via garden refuse (e.g. elephant-ears and yellow archangel). The acidity and water-holding capacity of the substrate, disturbance by fires, trampling and other human activities also determine the plant communities.

Reviewing the colonisation of artificial sites in Central Europe, Prach and Pyšek (1994) found clonal plants (those which self-replicate) were dominant more often than non-clonal plants (27:14 species respectively). Of the clonal species, around half were grasses or similar species (sedges and rushes), while almost all of the remainder were perennial herbs. In comparison, the non-clonal dominants were mainly annual (or biennial) species with a few woody perennials. It appears that some clonal and many annual non-clonal species are early colonisers, with clonal species spreading more rapidly. Later communities may become dominated by shrubs and trees (non-clonal species) with an associated decline in clonal cover.

Similar patterns have been recognised in Britain. Gilbert (1989 and 1992a) describes four stages of succession, beginning with those dominated by initial colonisers, leading to more mature communities. Examples of the communities found at each stage are shown in Table 2.2, although the actual composition depends on the site conditions. The first stage is the 'Oxford ragwort stage', where many of the species present produce abundant light, wind-dispersed seeds, and are typical of sites with intermittent disturbance and low competition. As sites develop, some annual species are replaced by perennials, and plants with a rosette structure decline while tall herbs such

Table 2.2:   Stages in the succession of plant communities on urban commons

Stage	Typical species	
The Oxford ragwort stage	Grasses	Annual meadow-grass, perennial rye-grass, creeping bent, Yorkshire-fog, barren brome, smooth meadow-grass, rough meadow-grass
From 0–3 years	Herbs	Colt's-foot, groundsel, knotgrass, Oxford ragwort, American willowherb, rosebay willowherb, mugwort, wormwood, white clover, dandelion, smooth sow-thistle, prickly sow-thistle, confused Michaelmas-daisy, common orache, tall melilot, weld, fat-hen, hedge mustard
	Woody plants	Goat willow, butterfly-bush
The tall-herb stage	Grasses	False oat-grass, cock's-foot
After 4–6 years	Herbs	Rosebay willowherb, confused Michaelmas-daisy, golden-rod, wormwood, fennel, goat's-rue, garden lupin, feverfew, tansy, Shasta daisy, melilot species, yarrow, creeping thistle, spear thistle, great willowherb, red clover, goat's-beard, mugwort, common mallow, buttercups (e.g. creeping buttercup), perforate St John's-wort, Jacob's ladder, common nettle
	Woody plants	Goat willow, butterfly-bush, grey willow, birch, eared willow, sycamore, brambles
The grassland stage	Grasses	Oat-grasses, common couch, cock's-foot, red fescue, Yorkshire-fog
After 8–10 years	Herbs	Tansy, confused Michaelmas-daisy, golden-rod, Japanese knotweed, red clover, creeping thistle, yarrow, large bindweed, ribwort plantain, greater plantain, cowslip, sheep's sorrel, curled dock
	Woody plants	Goat willow, butterfly-bush, grey willow, birch, eared willow, sycamore, brambles
The scrub woodland stage	Grasses	Similar to grassland stage
	Herbs	Similar to grassland stage
After about 10 years	Woody plants	Goat willow, butterfly-bush, common willow, eared willow, birch, ash, sycamore, broom, laburnum, rowan, hawthorn, elder, apple, Swedish whitebeam, brambles

Sources: Clarkson and Garland (1988); Gilbert (1989 and 1992a)

as rosebay willowherb dominate (see Species Box 2.9). In time, grasses represent a larger proportion of the total species, the tall herb community being replaced by grassland sward spotted with clumps of herbs. Many of the herbaceous species which do remain, have creeping stolons or spread vegetatively from rosettes in gaps within the sward.

....

Species Box 2.9: Rosebay willowherb

Rosebay willowherb is a tall perennial plant growing up to 1.5 m and producing pink-purple flowers from July to September. The leaves are long and thin, arising alternately from the stem. It is widespread throughout Britain and often abundant in woodland clearings and margins, scrub, embankments, demolition sites, quarry and mine waste, rubble and spoil heaps, walls, cliffs, waste ground and river banks. It is mainly found in slightly acidic soils, although it can also occur on base-rich substrates in demolition sites. It is usually absent from waterlogged soils. Although probably native to the uplands, it was introduced into lowland England from North America in the seventeenth century. The major spread was later, probably assisted by the railway network at the end of the nineteenth century, although it was also grown in gardens at this time. The abundant seeds germinate in bare soil, often soon after disturbance. It is also known as fireweed or bombweed because it is frequently found at fire-damaged and, during the last world war, bomb-damaged areas. Many insects use it as a food plant. These range from aphids and other bugs, to moth larvae (including some hawk moths), leaf beetles and flies.

Sources: Stace (1997); Grime *et al.* (1988); Rackham (1986)

....

Many woody plants establishing on demolition sites colonise within the first few years. After that, the herbaceous vegetation becomes too dense and saplings do not establish except where gaps open up. The earliest colonisers are those such as willows, birch, sycamore and butterfly-bushes (Species Box 2.10) which produce large amounts of wind-borne seeds and fruits. As the site matures, ash and hawthorn may establish, followed eventually by oak and beech.

Mosses and lichens grow in gaps or on prominent pieces of rubble, where the substrate type is important in determining the species present (Gilbert, 1988). Many of the common lichens are ruderal species such as *Peltigera didactyla*. Mortar-encrusted bricks house several species. Brick is rich in phosphorus, potassium, calcium and magnesium and supports lichens such as *Lecania erysibe*, *Scoliciosporum umbrinum* and several species of *Trapelia*. Common urban calcicoles (e.g. *Candelariella aurella*, *Caloplaca citrina* and *Lecanora dispersa*) are often found on lime-rich mortar. Other species occur on abandoned building materials (e.g. timber and concrete) and on rubbish (e.g. textiles).

In the early stages of colonisation by animals, the majority are those with good dispersal abilities such as birds, butterflies, moths, true bugs, some beetles and spiders. These are often clustered around those plants which have managed to establish. The majority of dispersing animals are winged; however, some small beetles and spiders are carried passively on the wind. Many spider species disperse by producing a line of silk attached to their abdomen which drifts in the wind

....

Species Box 2.10: Butterfly-bush

Butterfly-bush was first introduced into Britain around 1890, although it only really established from later stock brought in during the early 1900s. A shrub up to 5 m tall, it often occurs in waste ground, and on walls and buildings, sometimes producing quite dense thickets. The long, narrow, grey-green leaves have hairy under-sides which help to prevent water loss. This adaptation to conditions in its original home in central China also helps in dry urban habitats. It has long arching branches ending in pyramidal inflorescences which are usually lilac, purple or white and appear from June to September. The seeds are light, winged and wind dispersed. The dense flowers attract a wide range of insects, especially butterflies (hence its com-mon name). Not only do butterflies feed on the nectar, but its leaves provide food for the cater-pillars of at least eleven species of moths and butterflies as well as some beetles and bugs. A range of predators (e.g. spiders and ladybirds) can be found exploiting the herbivorous fauna.

Sources: Stace (1997); Gilbert (1989)

....

carrying the spider with it. This behaviour is known as ballooning and occurs especially in young spiders.

As the site ages and develops, there may be an increase in decomposers including earth-worms and woodlice (e.g. *Oniscus asellus*, *Androniscus dentiger* and *Trichoniscus pusil-lus*) and herbivorous species including the slug *Deroceras reticulatum*, the meadow grass-hopper and the field grasshopper (Gilbert, 1989). Other common animals, strongly associated with particular plants such as goat willow and butterfly-bushes, are butterflies (e.g. meadow brown, wall brown, common blue and small copper), skippers (e.g. small skipper and large skipper) and moths (e.g. six-spot burnet, the cinnabar, silver Y and ele-phant hawkmoth). Fifty-seven species of hoverfly were recorded from demolition sites

in Sheffield (Whiteley, 1988). Of these, twelve were found in over a quarter, and four (*Sphaerophoria scripta*, *Episyrphus balteatus*, *Syrphus ribesii* and *Eristalis tenax*), in over half of the forty-nine sites surveyed. In sixteen sites in Merseyside and Lancashire, 135 spe-cies of Hemiptera were found, including some new records for the area (Sanderson, 1992). More species and individuals were found in older, grass or shrub-dominated sites.

Predators might be expected to be among the last animals to arrive at a site, since they need other animals to feed on. This is true of some groups, especially those which require vegetation in which to wait for prey (orb-web-spinning spiders and some crab spiders). However, many active hunters (wolf spiders, money spiders, centipedes, harvestmen, rove beetles and ground beetles) are found even on bare ground. These animals, as well as feeding on each other, consume flying insects (especially flies) when they alight on exposed surfaces. For example, the Devil's coach horse beetle (Britain's largest rove beetle) may hunt for prey during the day on open areas between patches of vegetation.

As tall-herb and grassland communities develop, field voles may colonise and birds such as house sparrows, starlings, feral pigeons, goldfinches and kestrels use the site. Some large areas have nesting skylarks, meadow pipits and red-legged partridges (Gilbert, 1989). The development of scrub and woodland encourages a different fauna, with wood mice and bank voles being associated with areas of bramble, while more cover encourages wrens, linnets and dunnocks to nest.

Industrial Wasteland, Spoil Heaps, Quarries and Mines

The distribution of derelict land in England in 1982 showed a marked concentration in the north and west, much of which is linked to the mining and associated industries in these regions (Bradshaw, 1989). As a result, many derelict sites are in close proximity to residential areas. Many of these sites are contaminated by toxins and some may become eyesores (Plate 2.5). The three types of urban wasteland associated with industry are defined in Table 2.3. In Britain, by the mid-1970s, the Government acknowledged that there was over 71,000 ha of derelict land, to which could be added at least 268,000 ha of neglected land and substantial areas of operational land, for

Plate 2.5: Colliery spoil heap

Table 2.3:   Definitions of industrial wasteland

Derelict land	Land so damaged by industrial or other development that it requires treatment prior to beneficial use
Neglected land	Land capable of some beneficial use, but currently uncared for and detrimental to the environment
Operational land	Land within a current development which is detrimental to the environment and could be improved

Source: Bradshaw (1989)

which there are no figures. The area of derelict land increased in the period 1974–1982, despite reclamation taking place during the same period (Bradshaw, 1989). However, the presence of unusual habitats, interruption of successional processes and conditions of nutrient and water stress in industrial sites provide habitats for unusual communities (Gemmell, 1977; Bradshaw and Chadwick, 1980).

Industrial Wasteland, Spoil Heaps and Waste Tips

Land abandoned by industry is frequently low in quality and often toxic to plants. Poor physical structure of the substrate increases waterlogging and compaction in some sites, and instability, rapid drainage and erosion on others. Extremes of pH, low nutrient status, a lack of initial colonisers and toxic materials prevents rapid establishment of vegetation. Often lack of knowledge regarding the precise history of the site makes restoration more difficult. However, even fairly adverse and artificial conditions produce interesting habitats. There is no calcareous grassland in Greater Manchester, so it is a bonus to find calcareous plants, including many orchids, growing on tips left from the Leblanc process of producing sodium carbonate. Several regionally rare and local species of flowering plant are recorded from alkaline derelict land in southern and western Lancashire (Greenwood and Gemmell, 1978). These include colonies of marsh-orchids, two species of which (northern and southern marsh-orchids) are at the edges of their ranges. Other species found on inland sites, including yellow-wort and early marsh-orchid, are usually associated with coastal communities on sand dunes, dune slacks and saltmarshes. Pulverised fuel ash from coal-burning power stations has a high salt content

making it difficult for many plants to establish. However, where proximity to coastal seed sources allow, saltmarsh species colonise the early stages before the salts leach away (Shaw, 1994). Subsequently these sites are colonised by non-coastal species.

Other species are tolerant of toxicity (e.g. some grasses grow in the presence of heavy metal contamination) or pH (e.g. calcicoles grow on alkaline soils and acid-heath grasses dominate acidic colliery spoil). Plants such as dandelions (Species Box 2.11) often colonise exposed substrates on wasteland, where their deep tap-roots enable growth when nutrients and water are below surface layers of larger materials (e.g. stones and rubble). They often grow in dry soils, where the depth of the tap-root may be important in surviving summer droughts. In many sites as conditions change with age, scrub and tree growth develops (e.g. ash, birch, goat willow, hawthorn and sycamore).

The invertebrate communities of post-industrial sites, especially on disturbed (usually bare) areas, are often regionally uncommon, and in some cases contain notable species. For example, rich ground beetle faunas have been found in artificial sites (Eversham et al., 1996). Naturally occurring habitats of this sort are rare, many having been exploited or destroyed, and these similar derelict areas could be relatively easily maintained by a low level of disturbance (Eyre and Luff, 1995). However, conservation may be difficult since the prime reason would be for invertebrates (not generally high on the public agenda) and such sites are rarely aesthetically pleasing.

Quarries and Mines

Quarrying is a major industry in the UK. In 1988, some 67,270 ha of land in England were affected by surface mineral workings, not

Species Box 2.11: Dandelion

The dandelion genus comprises a large number of micro-species which are difficult to separate. Taxonomically they are divided into sections (nine in Britain), some of which are common in fertile and disturbed conditions, including gardens (e.g. sect. *Taraxacum*), while others are more likely to be found in drier habitats such as derelict land (e.g. sect. *Erythrosperma*). Some of the micro-species in each section are native, whilst others have been introduced. Dandelions are extremely effective at colonising new sites, especially exposed soil in disturbed areas. They form rosettes of lobed leaves close to the ground (less than 350 mm) with long tap roots. They produce yellow flowers (up to 400 mm high) from April until October (most before June) and set seed mainly in May and June. In most species the seeds are female, having been produced asexually. Seeds are numerous (up to 2,000 per plant) and attached to a downy extension which assists wind dispersal. Dandelions are perennials and overwinter as small rosettes of leaves. Regeneration may occur from fragments of the

tap roots, and large plants can produce several rosettes.

Sources: Stace (1997); Grime *et al.* (1988)

including sand and gravel pits, while over 7,800 km² were devoted to underground mining (Brown, 1992). About half of all derelict land recorded in 1988 resulted from mineral extraction. Of this, nearly 70 per cent is capable of being reclaimed, and in the six years before the survey 8,650 ha were reclaimed to forestry, agriculture, amenity, recreation, and for industrial uses. The resulting habitats present both problems and opportunities for wildlife (Table 2.4).

Abandoned quarries are slowly colonised from neighbouring areas and sometimes produce interesting communities; several disused quarries are now SSSIs (Humphries, 1980). Plant diversity may be related to quarry location, both geographically and in relation to the surrounding land use and vegetation. Another

factor influencing communities in quarries is quarry age, and hence the time available for colonisation. Quarry age also reflects changes in the extraction techniques employed. The best quarries for plant diversity and uncommon species, tend to be over forty years old (Davis, 1979). There are several studies on the relationship between quarry size and the number and range of species (e.g. Davis, 1979; Cullen, 1995). A direct relationship has not been identified, possibly because the diversity of materials within a quarry is more important in determining the species composition (Usher, 1977). In some quarries (such as those used for limestone extraction), excessive drainage, thin soils and low nutrient status promote successful colonisation of plants typical of species-rich grasslands (Table 2.5). Here, vigorous

Table 2.4: Problems and opportunities for wildlife arising out of mineral extraction

Factor	Problems	Opportunities
Habitat alteration	Direct habitat destruction including the removal of important habitats such as species-rich grassland and limestone pavement	Creation of new and varied habitats, both during working (e.g. walls, spoil heaps, lagoons) and after abandonment (e.g. lakes and ponds due to subsidence, quarry faces)
Pollution	Pollution can cause siltation of streams and rivers (e.g. from clay works), decline in aquatic species (e.g. from lead washings) and reduce the chances of vegetation colonising spoil heaps (e.g. from copper mine waste)	Some spoil heaps encourage hitherto local or even rare plant species (e.g. on lead mine waste or calcareous wastes)
Waste materials	Local environments can be contaminated by deposits of dust waste from limestone quarrying	Settling lagoons of calcareous quarry slurry waste mixed with brine provide opportunities for saltmarsh and dune slack vegetation
Quarry faces	High sheer, rock faces produced by large modern production quarries can take a long time to erode following abandonment	Rock faces provide nesting sites for cliff-dwelling birds such as the peregrine falcons, ravens, choughs, kestrels, jackdaws, stock doves, ring ouzels and wrens
Tunnelling	Extraction of stone by tunnelling can lead to instability of some areas of quarries	Tunnels left after extraction provides habitats for bats including local (e.g. greater horseshoe) and more common species (such as the lesser horseshoe, Bechstein's, Daubenton's, Natterer's and whiskered bats)
Soil removal	Large areas of land stripped of soil may lead to persistent scars on the landscape	Shallow substrates lacking nutrients and with poor water-holding capacities may encourage species-rich plant communities
Future developments	Future expansion threatens many habitats and their associated species	Planned restoration and reclamation provides opportunities for extending the range of habitats in an area following extraction

Sources: Ratcliffe (1974) with supplementary material from Bradshaw and Chadwick (1980) and Davis (1982b)

Table 2.5: Typical plants of chalk and limestone quarries

Grasses and sedges	*Herbaceous plants*		*Woody plants*
Cock's-foot (C, D, H)	Black medick (D)	Hogweed (C)	Ash (D)
Common bent (H)	Cat's ear (C)	Lady's bedstraw (C)	Bramble (D)
Crested dog's-tail (C)	Colt's-foot (C, D, H)	Mouse-ear-hawkweed (C, D)	Goat willow (D)
False brome (D)	Common bird's-foot-trefoil (C, D)	Oxeye daisy (C, D)	Hawthorn (D)
False oat-grass (C, D, H)	Common centaury (D)	Ploughman's-spikenard (D)	
Fern-grass (D)	Common dog-violet (C)	Red clover (C, D)	
Glaucous sedge (C)	Common knapweed (D)	Ribwort plantain (C, D)	
Quaking-grass (C)	Common mouse-ear (C, D)	Rosebay willowherb (C, D, H)	
Red fescue (C, D, H)	Common ragwort (C, D, H)	Rough hawkbit (D, H)	
Sheep's-fescue (D)	Creeping buttercup (C, D)	Selfheal (C, D)	
Smooth meadow-grass (C, H)	Creeping thistle (C)	Smooth hawk's-beard (H)	
Yellow oat-grass (D)	Daisy (C, D)	Tufted vetch (C)	
Yorkshire-fog (C, D, H)	Dandelion (C, D, H)	Wild strawberry (D)	
	Eyebright species (C, D)	White clover (C)	
	Fairy flax (C, D)	Yarrow (C, D)	
	Hawkweed species (C, D, H)		

Note and Sources: The species listed were found in 50 per cent or more of the quarries surveyed by Davis (1982b) throughout England (D), Hodgson (1982) in the Sheffield area (H), and Cullen *et al.* (1998) in Derbyshire (C)

species are unable to dominate and lower-growing, less competitive plants persist past the point at which they would be suppressed in more fertile environments.

Other mine workings (such as coal and metal mining) leave behind toxic waste tips which are only slowly colonised by plants and subsequently animals. Run-off from these may contain toxins and adversely affect local water courses. In contrast, opportunities for aquatic wildlife are created when subsidence of land once worked for coal extraction produces open water and marshland habitats. In Wigan, several such flashes (including Ince, Lightshaw Hall and Pennington Flashes) are fine examples of this. In addition, some spoil provides opportunities for any species tolerant to the contamination. The spoil from old lead mines, for example, encourages plants such as lady's bedstraw, bladder campion, common bird's-foot-trefoil and mountain pansy.

Refuse Tips

The growth of urban areas has inevitably been accompanied by a parallel growth in waste generation. Within densely inhabited areas, waste disposal (especially that of domestic refuse) is problematic. The method most commonly used is landfill (wet waste or sewage and industrial wastes are considered elsewhere). Landfill as a controlled method of managing domestic refuse was introduced into Britain in 1912, with the intention of allowing most material to degrade naturally, leaving the rest buried underground. This is space consuming, and in the past mainly utilised land considered inappropriate for other types of development. However, conflicts arise since much of this land (such as worked-out clay and gravel pits) has potential for creative conservation and management for recreation and wildlife.

Potential colonisers of refuse tips need to overcome, or be tolerant of, several problems (see Figure 2.6). Both heat and methane are produced during the degradation of organic material by micro-organisms. Methane and other pollutants leach out, creating toxic conditions. Water pools on the surface forming temporary ponds and accumulates in organic layers of refuse, producing waterlogged areas. Refuse settles with time, and steep slopes are unstable and vulnerable to erosion by wind and rain. As a consequence of this, and disturbance caused by tipping, many early colonising communities are temporary. Even once abandoned, and after a deep layer of top soil has been added (in Britain at least 1 m is specified), some problems persist, the main ones being methane production, toxic leachates and subsidence.

Many flying insects such as bees, wasps (especially the German wasp) and flies build nests, lay eggs and feed on rubbish. Some exploit the least likely of artificial items; the nests of rose-leaf cutter bees (*Megachile willughbiella*) and mason Bees (*Osmia rufa*) occur in items such as abandoned car batteries (Darlington, 1969). The warmth of decomposition provides suitable conditions for some species which are otherwise rare outside buildings in Britain: house crickets and cockroaches thrive in tips especially in the south. Mammals which utilise refuse tips include house mice, which may overwinter in the relative warmth, and foxes which forage amongst the garbage.

Ideally, little refuse should be exposed at any time and all should be covered within twenty-four hours with at least 0.2 m of soil or other suitable substrate (waste containing mainly biodegradable material requires at least 0.6 m). These guidelines are not always strictly adhered to. Uncovered rubbish provides food (especially in winter) for birds such as starlings and gulls. In Britain, frequent visitors to tips (even those many miles from the coast) are

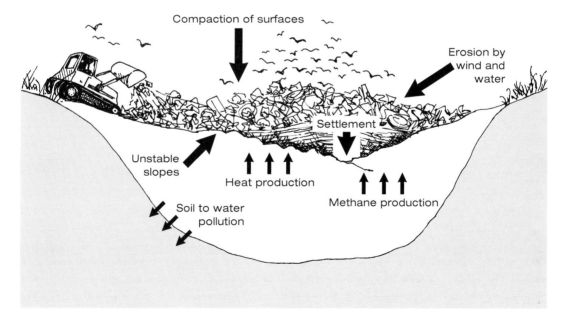

Figure 2.6: Problems for species colonising refuse tips.
Sources: Darlington (1969); Bradshaw & Chadwick (1980)

Species Box 2.12: Black-headed gull

The black-headed gull is the smallest common breeding gull in Britain (380 mm in length and 250 g). Adults have red bills and legs and dark brown hoods which end above the nape. The hood is lost in winter with only a small black patch remaining behind each eye. The leading edge of the top of the wing is white. As in many gulls, young birds have different plumage to the adults. The dark spots on the cheek and the whitish leading edge to the upper surface of the wings are the best features for identifying the otherwise brownish-grey immatures which have yellowish bills and legs. It is a very common species in both coastal habitats and inland waters, tending to prefer lowland areas (below 700 m). Black-headed gulls nest in colonies and sometimes with other species of gulls or terns. They breed at the margins of shallow, calm waters, usually on the ground in low vegetation. The nest is built mainly by the male and both parents share incubation and feeding the young. They lay one brood per year in May/June with replacement if the first one fails. There are usually three eggs per clutch, which are incubated for three to four weeks. Fledging takes place five to six weeks after hatching. Their food includes invertebrates, seeds, grain and small fish, often supplemented by refuse from rubbish tips in urban areas.

Sources: Cramp (1983); Harrison (1983)

black-headed (Species Box 2.12), lesser black-backed, herring and common gulls. This exploitation began early in the history of landfill, with records of gulls feeding at refuse tips from the 1920s. There are temporal and regional variations in the species found. In south-eastern England, black-headed gulls occur throughout the winter, common gulls are more frequent during February and March, and herring gulls more common during October and November (Bradshaw and Chadwick, 1980). In Scotland, black-headed gulls are less frequent on inland tips than lesser black-backed gulls (Furness and Monaghan, 1987). Large flocks occur at tips and feeding is highly competitive with larger individuals occupying the best areas. Near Sydney (Australia) over the last fifty years the human population has increased over four-fold while a local silver gull colony, which in the 1940s consisted of only a few pairs, is now over 50,000 pairs strong (Smith and Carlile, 1993). This increase in gulls is possibly due to the exploitation of relatively accessible food; within the area are one major and several smaller landfill sites, together with fish-processing factories and other urban food sources (such as restaurants, schools, etc.). In the Australian study, over 85 per cent of regurgitations from birds examined at their nests consisted of refuse (mainly meat with the remainder starchy material). Some gulls travelled up to 12 km to feed at the landfill sites.

Other habitats at active landfill sites include the cover material and standing water where drainage is impeded. Colonisation occurs at any area no longer disturbed by tipping. The species present vary with biodegradation and tip maturity (Darlington, 1969; Bradshaw and Chadwick, 1980). Where soils are damp, some early colonisers are nitrogen-fixing blue-green algae such as *Nostoc* and *Oscillatoria* species. Fungi including the shaggy ink cap, together with *Coprinus bisporus*, *C. cinerus* and *Volvariella speciosa* grow on organic material while *Lentinus lepideus*, *Tubaria furfuracea*, and *Dacrymyces deliquescens* (a gelatinous orange-yellow drop like fungus) live on rotting wood. Liverworts (e.g. *Lunularia crutiata*) and mosses thrive in damp patches, silvery thread-moss often being the commonest, with extinguisher-moss concentrated in calcareous areas on cement and mortar, and purple-fruiting heath-moss on more acidic substrates, often where fires have been. Annual species including annual meadow-grass and groundsel are usually the next to move in, followed by perennial species (e.g. docks and plantains) and eventually more woody plants. After as little as two years plant cover may become complete, with a gradual rise in the proportion of perennial species. Often, in areas still affected by disturbance from tipping, low-growing, rosette-forming plants are prevalent, for example the greater plantain, fiddle dock and dandelion. The rosette growth form protects plants from mechanical damage. Where disturbance is greatest, quite small plants grow in small depressions, only becoming large once the stress is removed. Some plants more often associated with cultivation occur, including oil-seed rape from agricultural waste, sweet-williams from garden refuse, tomatoes and potatoes from kitchen waste and cumin and coriander from bird seed. As time progresses, a rich soil fauna (including earthworms) assist in soil development.

Most landfill sites are covered with topsoil (or a mixture of subsoil and topsoil) after tipping is completed, and planted appropriately for the planned after-use. Those lying near to edges of conurbations are often used for recreation (golf courses, parks, etc.) and sometimes combined with management for wildlife. Post-tipping management often involves the use of impermeable linings to prevent leaching of polluted waters, capping (normally with clay) to reduce rainwater infiltration, and venting-off gas produced. As long as plants (especially trees) do not damage these caps or linings, then a range of habitats may be created, including wild-flower meadows and woodlands (although cover material needs to be deep to accommodate this). In some instances ponds are developed where the topography and waste settlement allows.

Road Verges and Railways

The sheer extent of the British transport network has major implications for wildlife. In the United Kingdom during the 1980s the volume of road traffic increased by 50 per cent and the road network grew by 5 per cent. Motorway traffic doubled and motorway length increased by 20 per cent to 3,100 km (Brown, 1992). This required the development of 14,070 ha of land (4,643 ha of previously urban land). There are approximately 11,000 miles of British railway lines and about 30,000 ha of adjacent land (Spellerberg and Gaywood, 1993). Roads and railways are both harmful and beneficial to wildlife. They are barriers to movement, and collisions with vehicles kills many animals. The development of transport routes leads to habitat loss, fragmentation and pollution stress. However, their verges may be diverse habitats (Colour Plates 5 and 6) and act as corridors for the dispersal of many organisms.

Road Verges

In 1965 there was a potential 2.5 million bird kills, with up to one-eighth of adult annual mortality in the house sparrow being due to roads (Hodson and Snow, 1965). However, for most species road mortality is much lower, and unlikely to affect population levels. Road kills of mammals can be high: of the eighty-seven fox cubs found dead in Bristol between 1978 and 1980, 59 per cent were road deaths (Harris, 1981). Attempts to address road mortality involve building tunnels to accommodate badgers, hedgehogs, toads and frogs. In addition, noise disturbance near roads may reduce densities of local breeding birds (Reijnen *et al.*, 1995). Roads act as barriers to some species, preventing safe passage between habitat patches for surface-active animals such

as mice and ground beetles (Mader *et al.*, 1990). Wide highways can have a similar barrier effect, for animals such as small mammals, to a water body twice the road's width (Oxley *et al.*, 1974). For other species, impacts of fragmentation, changes in topography, altered microclimate (including gusts from passing vehicles), effects on drainage and pollution from vehicles reduce population sizes. Species using verges as corridors risk traffic-related mortality and disperse more slowly along verges than in adjacent open areas (Vermeulen, 1994).

The most significant road-side pollutants are lead, nitrogen oxides and sulphur dioxide from exhausts and salt from de-icing in winter. Nitrogen oxides from cars consists of nitric oxide (NO) and nitrogen dioxide (NO_2), collectively known as NO_x. Levels of NO_x near some roads are in the range identified as poor air quality (100–299 ppb) by the Department of the Environment (Bower *et al.*, 1995). Both NO and NO_2 are taken up by plants through their leaves. Although some plants are intolerant to high levels of NO_x, in other species it is beneficial, stimulating amino acid production and hence growth. This increased growth has subsequently led to outbreaks of defoliating insects on plants on motorway verges (Port and Thompson, 1980). Even if NO_x alone is advantageous or neutral to a given plant, the accompanying sulphur dioxide is always detrimental (Dowdeswell, 1987). In Britain lead levels in petrol were reduced in 1985 and, although many cars now use lead-free petrol, some still need lead as an additive. Most of the lead emitted (70–90 per cent) is deposited as particles within about 50 m of the road and accumulates in taller vegetation. Lead reduces root and shoot growth, and may affect mineral nutrition, respiration and photosynthesis rates. Lead and cadmium from traffic was implicated in the reduction in fine roots and severe needle loss in Norway spruce along the sides of roads in Sweden (Majdi and Persson, 1989). However, plants of some species, for example ribwort plantain and red fescue, may develop tolerance when exposed to low levels of lead and cadmium.

Over a severe winter, motorway applications of common salt can be over 10 tonnes per lane kilometre. High salt concentrations break down the crumb structure of clay soils and produce an impermeable layer. Many plants are extremely susceptible to high salt levels. Salt remains within soil for a considerable time; the highest concentrations occur in spring and the lowest in October. The central reservation and the 2 m of road verge nearest to the carriageway are the worst affected. Salt pollution creates bare patches where plants die back (areas which are usually quickly recolonised each year). However, some salt-tolerant species (halophytes) such as reflexed saltmarsh-grass, sea plantain and lesser sea-spurrey are able to establish. These may grow some distance from their usual coastal habitats; their seeds probably being distributed by cars (Dowdeswell, 1987).

The road network provides a range of habitats, especially where there are substantial vegetated verges. A review of 212,220 ha of roadside habitat surveyed over ten years found approximately 870 (mostly native) species of plants, including thirty-five nationally rare ones (Way, 1977). For some, roadsides were important refuges, for example fen ragwort, thought to be extinct for many years, was rediscovered in a roadside ditch. Roadsides are important to many animals, with 40 per cent of mammal species, 20 per cent of bird species, all reptiles and almost all amphibians breeding in roadside habitats. Similar results were found for some invertebrates, with over 40 per cent of butterfly and bumblebee species breeding along roadsides. Up to twenty-three

species of butterflies and burnets occur at some roadsides (Munguira and Thomas, 1992). Large verges and the presence of nectar-producing plants are important for increased butterfly diversity, and whilst the level of traffic has little effect on dispersal or mortality for many species, wide roads may act as barriers to movement. Many of the ground beetles recorded on a sandy verge by Eversham and Telfer (1994) were from species known to be poor dispersers and/or were newly emerged animals, implying that they were residents rather than transient.

Amongst birds, the kestrel (Species Box 2.13) is one species frequently associated with roadsides and, although numbers have declined slightly recently, few motorway journeys are completed without sighting at least one kestrel hovering over the verge hunting for prey. They are the most common urban bird of prey, being adapted to a range of open urban habitats including grass verges along roads, railways, canals and rivers (Village, 1990). Their flexibility is mainly due to their use of a variety of nesting sites, diversity of prey and high hunting efficiency.

Railways

Following the decline of steam trains, the lower risk of accidental fire reduces the need to remove vegetation, although some management of active lines is required. The habitats associated with railway tracks are shown in Table 2.6. The permanent way is the strip upon which the track is laid, covered by a layer of cinders (the cess). Banks may be level (flats), built up (embankments) or dug into the surrounding landscape (cuttings). Drainage ditches, at the base of embankments, are lined with natural materials or concrete. Several structures, from bridges and buildings to yards and platforms complete the picture. Areas still

Species Box 2.13: Kestrel

In the kestrel, males are smaller (320 mm in length and 220 g) than females (350 mm in length and 230 g) and have bolder colouration. Males have grey heads and tails, spotted chestnut backs and black-spotted buff undersides, while females have rusty-brown heads and backs, with black bars on the back and tail. They have characteristic hovering behaviour, maintaining their position against the wind (kestrels cannot hover in totally still air). Kestrels use a variety of nest types (holes in trees, ledges on buildings, as well as the old nests of other birds and their own nests built in trees). Birds from northern Britain migrate south (to central and southern England and the Continent) in autumn, returning in early spring. Kestrels breed from March to June, producing one clutch of four to six eggs from mid-April to the end of May. A replacement clutch may be produced if the first fails. The male supplies the female with prey whilst she incubates the eggs (twenty-six to thirty-four days). Females initially brood the young while males bring prey to her. After about ten days he may help to feed the young, which fledge twenty-seven to thirty-two days after hatching. Most food (e.g. small mammals and insects) is captured on the ground, the bird either pouncing onto the prey from a perch or

from mid-air following its characteristic hovering behaviour. However, bird prey may be taken in flight. Larger food items may be cached for later use.

Sources: Village (1990); Cramp (1980)

in use are less diverse than abandoned tracks, stations and sidings, although even heavily travelled lines have some wildlife value.

In an extensive survey of British railway vegetation, sections differing in slope, age and surrounding landscape were found to have characteristic plant communities (Sargent, 1984). Although this study concentrated on rural areas, the issues of species spread and conservation are applicable to urban sections. Several species have spread along railway lines. For Oxford ragwort (Species Box 2.14) and rosebay willowherb, the network is likely to have been fundamental in determining present distributions (Rackham, 1986). Oxford ragwort was introduced into Britain in the late 1600s and with the development of the railways spread throughout urban areas. Rosebay willowherb is more widely distributed, and is frequent in urban and industrial sites. For both species, railways provide habitats and a distribution network for the light seed, which is often sucked along in the slipstream of trains (Rackham, 1986).

Spraying tracks with herbicide influences the species present. Winter annuals such as

Table 2.6:   Major railway habitat types

Habitat Type	Characteristics	Typical Plant Communities
Cess (cindery material between and along the tracks)	Free-draining, highly stressed, usually sprayed with herbicides	Few species grow where spraying occurs. In the absence of spraying, species present are typical of free-draining grassland. At the edges, species tolerant of disturbance (e.g. field bindweed, rosebay willowherb and field horsetail) may occur with some creeping species (e.g. bramble and ivy).
Track verges — Cuttings (positive slopes)	Slopes, often of natural materials, usually mineral soils, sometimes with exposed rock	The community depends on the inclination (loose substrates tend to lead to shallow slopes). If soils are thin, species-rich grasslands may develop (with species such as primrose and cowslip). With increased fertility, coarser grasslands eventually develop into scrub and woodland habitats (featuring lesser celandine and ivy). Grasslands are frequent on more regularly managed cuttings, while embankments often support scrub with woodland on the lower slopes where conditions allow. Steep sided rock, stone or concrete slopes may only feature low growing species such as mosses and lichens, unless cracks and crevices develop allowing ferns and higher plants to colonise.
Flats (level areas alongside the cess)	Level areas, often of natural materials	
Embankments (negative slopes)	Slopes, often built with introduced materials	
Drainage ditches (usually at the base of embankments)	Often dug into natural materials (although drainage channels at the base of cuttings are often concrete)	Usually typified by aquatic or wet-soil-loving species if natural substrates are present. Mosses may grow on concrete channels and some higher plants may be found if sediment accumulates (e.g. common reed, greater spearwort and branched bur-reed).
Yards and sidings	When in use may be highly disturbed, however are frequently abandoned and allowed to colonise	Those in use have similar communities to other railway habitats (platforms and track verges/cess respectively). If abandoned, then succession may take place in a similar way to that along disused tracks, producing fairly pure grass swards, which quickly convert to scrub (perhaps including wild privet and elder) and some woodland (silver birch and ash).
Masonry (buildings, platforms, tunnels, bridges and concrete posts)	Often harsh and lacking moisture	Provide habitats for species which are tolerant of the extreme conditions on concrete, rock or brick surfaces. Mosses and ferns occur where shaded, lichens on exposed surfaces, and some higher plants (e.g. broad-leaved everlasting-pea, large bindweed and butterfly-bush) may root into crevices and cracks.

Sources: Sargent (1984); Gilbert (1989); Wright and Wheater (1993)

common whitlowgrass and thale cress complete their life-cycles before spraying takes place, and survive dry summer conditions as dormant seeds. Others, like sticky groundsel, grow after spraying. Some that do not occur because their growth period coincides with the time of spraying, are increasingly common on stretches where spraying does not take place

---

Species Box 2.14: Oxford ragwort

Oxford ragwort is an erect plant (which may have woody lower parts) and grows to some 0.5 m in height. The leaves are long with deep lobes, the end one of which is especially pointed. It is a perennial plant, albeit short lived. Yellow flowers are produced from May to October, although flowering can occur for most of the year in some urban sites. From June, large numbers of plumed, wind-dispersed seeds (about 10,000 per plant per year) are produced which germinate in bare ground or gaps in vegetation. Oxford ragwort was originally introduced into the Oxford Botanical Garden in the late 1600s from Mount Etna, and was first noted on walls elsewhere in 1794. It is now characteristic of derelict urban sites, having extended its range dramatically. It is common in southern Britain and more local in the north. It appears to prefer soils of high pH and is absent from woodlands, wetlands and sites which are grazed or heavily mown.

Sources: Stace (1997); Grime *et al.* (1988); Rackham (1986)

(for example small toadflax and field bindweed). Changes since 1960 in the methods used to reduce vegetation height (i.e. cutting and burning are now less frequent) led to an increase in fine-leaved grasslands (Sargent, 1984). The length of track which is now disused amounts to over one third of that in use fifty years ago. Associated abandoned stations and sidings provide other habitats for wildlife. Disused tracks support quite different communities to those where trains still run (Plate 2.6), the resulting communities depending on the physical nature of the environment and the use to which the land is subsequently put. Several distinct plant communities were found on a disused section of line in Derbyshire, closed to trains for twenty-two years (Wright and Wheater, 1993). The species were influenced by the relief (steep cuttings being richer in species than embankments) and access (more species generally being in closed areas compared to those open to the public). On undisturbed shallow slopes, scrub and (in some cases) woodland had developed. On highly disturbed areas (where public use was frequent), only low-growing herbs survived well.

For many animals, tracks are barriers to dispersal in the same way as narrow roads. Arthropods such as ground beetles tend to move parallel to tracks rather than crossing the line (Mader *et al.*, 1990). For other species, railways, especially disused tracks, can assist in dispersal. In Greater Manchester, urban foxes use such corridors, spreading out into parks and gardens from points of access. A species of bug (*Rhytistylus proceps*), usually associated with coastal or calcareous habitats, thrives on the calcareous flora growing on

Plate 2.6: Overgrown disused railway line

limestone ballast on a disused railway cutting (Sanderson, 1992). The population had probably colonised over a considerable distance, the nearest coast being 20 km away.

Walls, Buildings and Paved Areas

Some of the harshest urban habitats are constructions of brick, concrete and stone which, at best, provide analogues of cliffs and bare rock. They are inhospitable, often deficient in moisture, nutrients and shelter, and yet may be rich in both plants and animals.

Walls

It has been estimated that there is 1 ha of vertical wall surface for every 10 ha of urban habitat (Darlington, 1981). Surveys of the Roman walls in Colchester recorded sixty-six species of higher plant (including the nationally scarce lesser calamint) and 125 species of lichens (including some which are nationally rare) and fungi (two of which were previously undescribed species). Many lichens were present in large numbers, including a previously undescribed species. In addition, the invertebrate fauna includes some nationally rare species including the ant *Ponera coarctata*, and the bug *Asiraca clavicornis*. It is not surprising that structures this old are colonised by so many species. However, even on modern walls where conditions are dramatically different, there are habitats for many species. New walls have steep, vertical, highly exposed faces with little nutrient material, although this is modified in time as the wall decays. Mortar usually erodes faster than the bulk of the wall, creating crevices which are exploited by plants. In time, organic material builds up in cracks, on ledges and other protruberances of the wall, and especially where lichens, mosses and climbing plants grow. This enables grasses, ferns and flowering plants to gain a foothold. These larger plants accelerate the erosion and deposition of organic material and provide shelter, allowing a more diverse community to develop.

Not only do wall floras vary across the

country, partially reflecting climate and partly as a result of the surrounding vegetation, but the species composition depends on the conditions in different sections of individual walls (see Table 2.7). Many species grow better where the wall is wetter; on north-facing sides and where there are deeper cracks or loose substrate. Chemical and physical properties of the substrate are important; walls built of limestone promote calcareous species, while hard bricks and concrete are smoother and deteriorate more slowly than stone. The dispersal methods used by plants also influence the species. For example, ants are regular scavengers on walls, and frequently carry seeds long distances back to their nests. Some seeds are more attractive to ants than others, being oily on the surface (e.g. snapdragon, wallflower, ivy-leaved toadflax and herb robert), and those lost along the way germinate in cracks in the wall (Gilbert, 1992b).

When examined closely, even the smallest plant communities on walls provide suitable conditions for animals. From 216 samples of mosses collected from walls near Zürich, over 60,000 animals from 194 taxa were found (Steiner, 1994). These included nematode worms (19,054 individuals from forty-seven taxa), tardigrades (8,636 from thirteen taxa), rotifers (6,052 individuals) and arthropods, of which mites and springtails were the most common (11,644 from seventy-four taxa). There was also a sprinkling of larger animals such as spiders, woodlice, centipedes, beetles and other insects. These larger arthropods, together with other invertebrates such as snails and slugs live amongst vegetation and within cracks on many walls. Snails prefer walls made of limestone, or with calcareous mortar, because they use the calcium for their shells. They are vulnerable to drying out and take shelter in cracks, venturing out when it is humid. Woodlice also shelter from dryness and often aggregate under flaking surfaces or

Table 2.7: Typical plant species (not including climbers or grasses) found on boundary walls

Position on wall	Conditions	Typical plants
Top	May be level but often very exposed. May be main site for seeds dropped by birds	Specialist plants, some woody species and some garden escapes including barberry, black nightshade, cotoneasters, flowering currant, gooseberry, polypody, red valerian, stonecrop species, wallflower
Middle to top	Vertical, highly stressed (lack of water and high exposure)	Specialist wall species include lichens and mosses together with black spleenwort, ivy-leaved toadflax, maidenhair spleenwort, rustyback, wall-rue, yellow corydalis
Ground to middle	Vertical, but may have some moisture from damp rising and soil and water splashes from the ground	Many weed species including clustered dock, creeping yellow-cress, dandelion, groundsel, gypsywort, herb robert, mind-your-own-business, procumbent pearlwort
Ground at base	Often fertile soils (high pH if the wall was built using lime mortar), wettest part of wall	Mosses and some liverworts where wet, together with many weed species and garden escapes including cleavers, common nettle, knotgrass, prickly sow-thistle, rosebay willowherb, snapdragon, wood avens

Sources: Darlington (1981); Gilbert (1992b)

in cracks. Walls are important habitats for many woodlice (Species Box 2.15), with sixteen of the thirty-seven non-alien British species having been recorded from walls with mortar, and several others from drystone walls (Harding and Sutton, 1985). Of these, the most frequent are *Oniscus asellus* (25 per cent of records), *Porcellio scaber* (23 per cent), *P. spinicornis* (14 per cent), *Philoscia muscorum* (10 per cent) and *Armadillidium vulgare* (7 per cent). With the exception of *P. spinicornis* these are four of the five commonest species in the British Isles. *P. spinicornis* is rarer elsewhere, possibly because it has a heavy exoskeleton (which contains calcium) and may prefer calcareous environments.

Spiders are frequent inhabitants of walls.

Some, such as the zebra spider *Salticus scenicus* (a jumping spider), frequent bare surfaces, especially in bright sunlight. This animal moves in jerks, jumping onto its prey of flies. It lays down drag lines of silk for crossing gaps and returning to the wall when jumping. Other spiders produce webs from projections on the wall and from plants. One common spider, *Segestria senoculata*, uses a different type of web, comprising a tubular retreat (built in a hole) with several radial lines extending from the opening. The spider waits at the entrance with its front legs resting on the web, dashing out when prey disturb the radiating threads. It then carries the prey to the retreat to feed. Other arachnids which frequent walls include harvestmen; nocturnal

....

Species Box 2.15: Woodlice

Woodlice are distinguished from other similar-looking animals by having seven pairs of legs and being flattened from top to bottom. There are thirty-seven species of woodlice which live in the wild in Britain (some of which are quite rare or have narrow habitat requirements) and another ten or so species which are restricted to greenhouses. The five most common species (*Oniscus asellus, Porcellio scaber, Trichoniscus pusillus, Philoscia muscorum* and *Armadillidium vulgare*) are present throughout the country in a range of habitats (although in the north of Britain *A. vulgare* tends to be more closely associated with the coast). In many places, large aggregations can be found. Woodlice tend to be vulnerable to drying out and are very common in damp places (e.g. under logs and stones, on walls, and in buildings), although some species may be found in the open in bright sunlight (e.g. on walls). They are important decomposers of dead and decaying plant material which they break down into small fragments that subsequently decompose very quickly. They also feed on hyphae (the filamentous growth of fungal tissue) growing on decaying vegetation, and very

occasionally eat living plants. After a simple courtship where the male taps the female with his antennae, he transfers sperm to the female. The eggs are carried by the female in a fluid-filled pouch on her underside. Juveniles emerge with six pairs of legs, but obtain their seventh pair very soon after. Many species of woodlice reach maturity after a year and live for a further year or more.

Sources: Oliver and Meecham (1993); Hopkin (1991)

....

hunters which often rest on walls in aggregations and move delicately across the surface in search of small invertebrates. Several Hymenoptera also live in and on walls. Honeybees (*Apis mellifera*), and more often bumblebees (e.g. *Bombus pratorum* and *B. pascuorum*), nest in wall spaces. However, it is mining bees (species of *Andrena* and *Halictus*) for which walls are a major habitat. These are semi-social or solitary and either use existing crevices or produce narrow tunnels into softer parts of walls within which they build their nests. Ants (especially the black garden ant and the yellow meadow ant) forage on walls, and occasionally build their nests in holes. Other species shelter in cracks at the base of walls, venturing out to feed (e.g. rove beetles such as the Devil's coach horse beetle). Animals such as butterflies and common lizards bask on the bare surfaces of walls.

Buildings

The external surfaces of buildings have similar conditions to boundary walls, although many plants do not grow high on buildings, and concrete does not provide the nooks and crannies associated with brick and stone. Derelict buildings are also quickly colonised (Plate 2.7). City buildings provide nesting, roosting and perching sites for birds, together with habitats for lichens, mosses, ferns, flowering

Plate 2.7: Trees growing inside a derelict city-centre church

plants and invertebrates. Window boxes may have limited spaces for wild plants and animals. Roof gardens provide some opportunities, although most are primarily intended for aesthetic and recreational purposes. Other features which encourage wildlife include the positioning of bat bricks and bird boxes within the design of the structure. Ledges encourage birds to perch, roost and nest. However, with species such as feral pigeons, herring gulls and (in winter) starlings, this may cause problems. Large densities of these species in some British cities in recent years has led to problems of noise and fouling of buildings and streets. Prevention is difficult, although encouraging birds of prey may help to control the population.

Several birds of prey are adapted to life in cities. In the early 1800s, peregrine falcons were documented from towns and cities across Europe, and later North America. In the late 1950s there was a major decline worldwide, associated with the use of organophosphate pesticides, by which time it rarely nested in urban areas. Populations rallied during the 1970s and peregrine falcons now reside in many western European and northern American cities, often as a result of reintroductions (Luniak, 1995). Their cliff-nesting habits and diverse prey make them particularly suitable for urban dwelling. In Canada, urban merlins have higher densities (up to 25 pairs km^{-2}) and lower egg failure rates (only 4 per cent of nests) than rural Canadian or British populations, which have densities of up to 19 pairs km^{-2} and egg failure rates of 30–97 per cent of nests (Sodhi *et al.* 1992). Merlins have increased in urban environments since an initial colonisation in the late 1960s and early 1970s. Urban areas have fewer predators, especially of eggs and nestlings (other birds of prey predate merlins elsewhere). More importantly, the major limiting factors in rural

areas (scarcity of prey and nesting sites) are probably not a problem in urban areas. Merlins do not build their own nests, therefore the movement of crow species into cities provides nest sites. Merlin prey includes house sparrows which are often abundant in cities. Urban kestrels usually nest in natural and artificial situations from above 1 m (but sometimes on the ground) up to 50 m high on buildings. In Bratislava, kestrel nests are mainly on artificial structures (26 per cent on buildings under construction), with trees used only in non-built up areas (Darolová, 1992). Nesting height was greater in the inner city than on the edge, and nest sites changed frequently between years, mainly due to destruction or disturbance of existing sites. Although kestrels usually feed on small mammals, in urban areas where these are relatively scarce they take more small birds. In one urban study (Yalden, 1980b), around 76 per cent of the prey by weight were birds (mainly passerines, especially house sparrows, with some pigeons, possibly fledglings), and 21 per cent were small mammals (mainly rats and house mice) with relatively little invertebrate food.

A wide range of birds occasionally nest on roofs in Britain (e.g. oystercatchers, common gulls, and ravens), while species such as herring gulls and lesser black-backed gulls are more frequent (Fisk, 1978). Although some species are opportunists, many are typically cliff-nesting species. This adaptation may be a forced one, due to increasing human influence on natural areas such as beaches and islands. In comparison with natural areas, gulls in towns have a lower nesting density, a higher breeding success, and an increased tendency to change nesting site the year after a poor breeding season. One bird intimately linked to buildings is the house martin (Species Box 2.16). In South Manchester, Tatner (1978) found an increase in the number of nests (by

··

Species Box 2.16: House martin

House martins are about 125 mm in length and 20 g. They have short forked tails, are white underneath and, except for a white rump patch, are entirely blue-black above. Juveniles have slightly brownish upperparts. They are the most urban of the Hirundinidae (martin and swallow family) and are often, although not exclusively, found near water. They frequently nest in small colonies and, in urban areas, build nests of mud incorporating soft plant material usually just under the eaves of buildings. They have one or two broods per year from the end of May to August. Four to five eggs are laid per clutch, which hatch after thirteen to nineteen days. The young fly some nineteen to twenty-five days after hatching but are fed by the parents for a further one or two weeks. Both parents build the nest, incubate the eggs and feed the young. House martins feed on insects (especially flies and aphids) caught on the wing, sometimes at high altitudes. House martins may form post-breeding

flocks from July to October prior to migration. They migrate south (to Africa) in September or October, returning the following spring or early summer.

Sources: Turner (1989); Cramp (1988)

··

71 per cent) and number of colonies (by 31 per cent) compared to a survey about 40 years previously. There was a particular increase in the number of larger colonies (above six nests per colony) and some indications of an increase in more urban areas. These increases may be due to a decline in some atmospheric pollutants following the introduction of smokeless areas within Manchester. Apart from direct benefits on the birds, it may reflect indirect effects on their insect prey.

Buildings and other aerial structures are hazardous to birds. In the USA, about 550,000 birds per day die as a result of collisions with artificial structures (Gill *et al.*, 1996). Collisions with power lines and other wires are probably increased by lack of visibility and collision levels are higher when lines are orientated across flight paths, for example near to take-off and landing sites for waterfowl and swans. Where this problem occurs, warning objects can be attached to the lines to increase their visibility.

Of all animals living within buildings, bats tend to be one of the most misunderstood by the general public. The Wildlife and Countryside Act (Department of the Environment, 1981: see Jones, 1991) protects all bats in Britain, and includes the necessity to seek permission to remove or otherwise interfere with bats in houses. Timber treatment has been a particular problem for bats in the past, with one of the most commonly used chemicals (lindane) being implicated in bat mortality (Mitchell-Jones *et al.*, 1989). Although use of alternatives such as pyrethroids should reduce mortality, the lack of public awareness regarding the illegality and toxicity of lindane may mean that applications of this highly persistent chemical are still taking place. Bats are

Species Box 2.17: Pipistrelle bat

Pipistrelle bats are the smallest (3–8 g; head and body length 35–45 mm; wingspan 190–250 mm) and most abundant British bat. They are dark brown above and lighter brown underneath with dark brown to black wings and short muzzles. Pipistrelles are very common in a wide range of habitats. Mating usually occurs in autumn, although (unusually for mammals) females store sperm until they ovulate in spring. Pipistrelles overwinter from September to March in hibernation sites in which both sexes and all ages roost together in small groups. Pipistrelles are inactive during this period, although they occasionally feed if conditions are appropriate (usually at temperatures over 8°C). In summer the sexes roost apart, the males singly or in small groups and females in large maternity colonies. Females move into these roosts from March and April, giving birth to single (occasionally twin) young in late June to mid-July. Maternity roosts can be huge (up to 1,000 bats) although smaller ones (less than 200 animals) are more common. The young fly when three weeks old. Young females may mate that autumn, although males are not mature until they are a full year old. Pipistrelles feed at night, emerging just after sunset to feed (on the wing) up to 5 km from the roost. The main prey are flies and caddis flies together with other flying insects (mayflies, lacewings and

moths). They usually make one feeding foray of about four hours per night, eating the prey in flight. However, before the young are weaned, females feed twice during the night (just after dusk and again before dawn), returning to the roost to suckle the young in between trips.

Sources: Corbet and Harris (1991)

generally more common in buildings in the south, with only the pipistrelle (Species Box 2.17) and brown long-eared bats at all frequent in the north (Mitchell-Jones, 1989). Although pipistrelle and brown long-eared bats are the most frequent in houses (54.5 and 33.7 per cent of 1,276 identifications made between 1982 and 1986 respectively), 10 other species were recorded at frequencies of 4 per cent or less (Mitchell-Jones, 1989). Within buildings, bats not only roost in roof spaces or under eaves but also in cellars; in general they are distributed in relation to temperature.

Other adaptations to urban life include feeding on insects attracted to urban lighting. In Sweden, bat species which fed around street lamps suffered less from a decline in insects within urban areas than those (such as some *Myotis* and *Plecotus* species) which did not (Rydell, 1992).

The inside of buildings provide suitable conditions for many non-native animals. Three mammals are commonly found in association with humans (i.e. are synanthropic): the house mouse which arrived in the Iron Age possibly from the Asian steppes; the ship rat, probably introduced from India in the third century; and the common rat which is a later arrival (early 1700s) from Eastern Europe. Many introduced invertebrates are associated with buildings, some as pests of food or other organic materials (such as textiles) and others simply living alongside people. The slug *Limax flavus*, which was introduced in 1936 (Gilbert, 1989), is common in cellars and outhouses, several species of ants and millipedes have become established in greenhouses, and introduced cockroaches (Species Box 2.18) are pests of houses, hospitals and factories. Native British cockroaches do not inhabit buildings; it is the introduced German, Oriental and American cockroaches (especially the former two) which are synanthropic.

Native species also find their way indoors; butterflies such as the small tortoiseshell (Species Box 2.19) and peacock overwinter in attics, and woodlice, some beetles and several spiders all find niches inside buildings. Spiders are very common in buildings; *Amaurobius ferox* and *Tegenaria duellica* are common in houses, whilst *Pholcus phalangioides* is common in older buildings, often spinning webs in the upper corners of rooms, large cupboards and bathrooms. In some cases occupation of houses is opportunistic (such as weevils living on house plants). In other species large numbers of individuals may invade buildings, perhaps in search of food (e.g. ants), damp conditions (e.g. woodlice) or as a result of disturbance to their habitat nearby (e.g. millipedes).

Paved Areas

Plants colonising pavements need to tolerate many stresses: salt from de-icing; pollutants from cars (which have been considered earlier); extremes of temperature; compaction and trampling (which are discussed here). Tramp-

ling not only causes mechanical damage to plants, but also increases compaction of the soil, leading to poor moisture retention and lack of aeration. Plants growing out on the pavement often have a rosette growth form (e.g. dandelions and plantains) or occur in low-growing mats (e.g. wall-screw moss and silvery thread-moss), both of which are adaptations that reduce damage from trampling. Grasses on pavements are low growing because the tops are continually damaged, while the leaves grow continuously from the base (replacing lost tissue). The most common pavement grass (annual meadow-grass) is highly tolerant to trampling and sets seed even when it is reduced to a very short plant. The gaps between paving stones accumulate moisture and nutrients, and where these spaces are large, plant communities resemble those of heavily trampled gravel paths. With high levels of nutrients, species composition is similar to that in field gateways and well-trodden paths, for example perennial rye-grass, annual meadow-grass, several plantains and shepherd's-purse. Where trampling is very heavy, lower-growing plants such as procumbent pearlwort and mosses occur. Populations are sparse, therefore plants tend to reproduce by self-pollination and produce abundant very small, wind-dispersed seeds (e.g. procumbent pearlwort). For some species, vehicles are a major dispersal medium; pineappleweed, an introduced species (first recorded from Britain in 1871) has spread throughout the country, probably by its tiny seeds being dispersed in mud attached to feet and vehicles. In the exposed conditions of pavements, annual species often survive winter weather better than perennials, except those with deep roots (e.g. dandelion), since dormant seeds cope better with freezing conditions than do whole active plants. Plants growing away from the kerb, and especially those

Species Box 2.18: Cockroaches

The common (or Oriental) cockroach and the German cockroach are the cockroaches most frequently found in buildings, although both also occur on rubbish tips in summer. The common cockroach prefers warmer surroundings and is less frequent than the German cockroach. The American cockroach is even rarer outside buildings, and tends to be restricted to warehouses and industrial premises. Despite their names, all three originated in Africa and were introduced into Britain. The common cockroach is 17–30 mm long and shiny black, with short-winged males and wingless females. The German cockroach is smaller (10–15 mm long), lighter brown and both sexes have long wings. The American cockroach is the largest in Britain (27–44 mm long), reddish-brown and is fully winged in both sexes. Those with wings tend to glide rather than fly, since most cockroaches are fast running, ground-living animals. They are active at night and are omnivores, causing damage to stored food and materials. Breeding occurs throughout the year, and a single female produces many eggs. As the eggs leave the oviduct they are stuck together in symmetrical double rows to form an eggcase (ootheca). The ootheca takes several days to produce and the size and shape is species specific (8×3 mm, light brown containing thirty to forty eggs in the German cockroach; 10×5 mm, dark brown containing about fifteen eggs in the common cockroach; and 8×5 mm, dark brown containing fifteen to twenty-eight eggs in the American cockroach). Female common and American cockroaches deposit the ootheca long before the nymphs hatch, whilst in the German cockroach it is attached to the female until hatching occurs. The time until hatching differs between species, with the German cockroach having the shortest period (twenty to thirty-five days), and the other two species taking almost twice as long (common cockroach, forty-five to fifty-six days and American cockroach, forty to forty-nine days). The nymphs (which look like wingless adults) develop into reproductively mature animals in one to two months for German, five to seven months for common, and six to twelve months for American cockroaches.

Sources: Marshall and Haes (1988); Bell (1981)

••

Species Box 2.19: Small tortoiseshell butterfly

Small tortoiseshell butterflies have reddish upperwings with black and yellow markings and blue spots around the edges of the wings. Males are slightly smaller (wingspan 45–55 mm) than females (wingspan 52–62 mm). Eggs are laid in clusters of about eighty on the underside of the subapical leaves of nettles at the end of April. Larvae hatch in May and feed gregariously on nettles, and are yellow and black in colour, advertising the fact that they contain toxins. They use silk to connect the top leaves of the nettle together, producing a communal web in which they live until after their final moult, and finally disperse from the nettles to pupate suspended in shrubs, hedges or on walls. Adults emerge in June. A second brood often occurs in midsummer (with eggs present during July and August), especially in the south. Adults from the second brood hibernate over the winter, emerging in March or April to look for mates. Adults from a first brood that has been delayed by a late spring (especially further north) may also hibernate, saving their eggs until the following spring.

Small tortoiseshell butterflies often hibernate in houses and other buildings, and sometimes wake up in autumn when heating is switched on. Adults feed on nectar, and often feed in gardens.

Sources: Emmet and Heath (1989); Carter and Hargreaves (1986)

••

near to walls, are sheltered from many of the problems associated with pavements, and may receive additional nutrients washed from walls by rain (Plate 2.8). Species common at the base of walls include sow-thistles (both prickly and smooth) and rosebay willowherb.

Animals such as dogs, cats and feral pigeons are common on paved areas. Dog urine and faeces affects the urban environment, for example by providing nitrogen which may lead to plant communities typical of field gates (see above). The influence of dog urine on algal and lichen communities can be seen at the base of trees where urination is often concentrated. Here, dark green patches (especially when wet) of the alga *Prasiola crispa* together with other algae, lichens such as *Lecanora dispersa* and the moss *Bryum capillare* contrast with the pale green lichen *Lecanora coniza-*

eoides, or the bright green alga *Desmococcus olivaceus* on the rest of the trunk (Gilbert, 1989). This is especially obvious on isolated trees at the entrances to housing areas or greenspace. Pet cats and dogs may influence urban habitats by taking small amounts of prey and depositing waste materials. Probably of more impact are uncontrolled feral populations of both species. Feral cats, in particular, may live in large urban populations, sometimes exceeding levels which can exist on the available food, the extra often being supplied by householders. The availability of shelter is more important in some areas than supplementary feeding in maintaining large populations (Calhoon and Haspel, 1989). Feral pigeons (Species Box 2.20) are one of the most ubiquitous urban species, living in habitats as diverse as suburban gardens and busy city

Plate 2.8: Vegetation growing at the base of a wall

....

Species Box 2.20: Feral pigeon

The feral pigeon is derived from the rock dove (probably from domestic stock kept for food). Domestication has taken place since neolithic times and pigeons were established in towns by the fourteenth century, although they did not become feral until much later. They are around 330 mm in length (260 g) and very variable in colour, although many are mainly grey with shiny green and lilac patches on the sides of the neck and two black bands on the rear half of the wings. They have white underwings which are most obvious in flight. Albino, dark, chestnut and blue-chequer (blue-grey speckled with black) morphs are all common. Although feral pigeons breed all year round, at any one time perhaps only a third of the population is actually breeding, and they often stop during late August and September (the main period of moulting). They often breed in groups, with nests in positions inaccessible to cats and sheltered from direct rain. They lay two eggs per clutch which are incubated for seventeen to nineteen days and the young take four or five weeks before fledging. They feed in flocks in sites including rubbish tips, town centres and grain stores. They are opportunistic feeders of refuse, food deliberately given to them, grain, as well as natural seeds and fruits. Feral pigeons are fairly sedentary; feeding, nesting and roosting sites are often within several hundred metres of each other.

Sources: Ewins and Bazely (1995); Cramp (1985)

....

streets. When undisturbed, they have small home ranges, little mixing between groups and low individual mobility (Sol and Senar, 1995), probably because feeding, roosting and breeding areas are close together. However, replacement is fast if populations are removed.

Churchyards and Cemeteries

There are over 20,000 churchyards in England covering around 10,000 ha (Dennis, 1993). Many were originally grasslands enclosed either when the church was built or as a later extension. They range from manicured lawns and tended flower beds to neglected wildernesses (Plate 2.9). Even in the former, the ground between graves may have been undisturbed for some time except for mowing or (in more rural areas) grazing. Since many churchyards originate from grasslands, it is not surprising they are often refuges for meadow plants such as meadow saxifrage, cowslip, oxeye daisy, lady's bedstraw, pignut and burnet-saxifrage. Over 100 species of plant

may occur in an average-sized churchyard (about 0.5 ha). Typically older churchyards have more native species, with mature yew (Species Box 2.21) and beech dominating, interspersed with lime and exotic conifers (including monkey-puzzle and cypresses) which were often planted in Victorian times. Shrubs including holly and climbers like ivy are also typical.

Church buildings provide nesting sites for jackdaws, swifts and barn owls, and bats sometimes roost in the roofs or walls. The most common bat is the pipistrelle, although brown long-eared bats are also frequent, and Natter's and Daubenton's bats have been recorded from churches in Northamptonshire (Dennis, 1993). Church porches are common roost sites, especially those which are south-facing and frequently warmed by the sun. The surrounding land influences the animals taking shelter or food; in suburbia many species are inhabitants of gardens, while on the urban fringe they are more commonly those associated with agricultural habitats. Small mammals, foxes and occasionally badgers also

Plate 2.9: Heavily vegetated urban cemetery

Species Box 2.21: Yew

Yew is a native, evergreen conifer with reddish-brown bark and grows up to 28 m, although many in urban settings do not exceed 15 m. It often has several trunks and, in urban situations, is sometimes planted as a hedging tree. Yew has a local distribution, being rare in Scotland, but more common in England and Wales. The dark green needle-like leaves are arranged in two rows. Yew is dioecious (male and female flowers grow on different trees) with small (1–2 mm) green female flowers and larger (5–6 mm) male flowers which appear yellow due to a covering of pollen. The fruits are red, fleshy and berry-like, around 8–12 mm in size. Both the leaves and fruit are poisonous to cattle and humans (although not to deer or birds which feed on the berries). Its toxic nature has been suggested as a reason why it is often found in churchyards, since it was planted (to produce longbows) but needed to be kept away from grazing livestock. Growth in diameter can be very slow (sometimes only 10 mm in twenty years) and the yew is the longest living tree in northern Europe. It is shade tolerant and casts a heavy shade itself, reducing competition from other trees and shrubs. Few insects live on yew, those that do include three moth, two bug and one fly species.

Sources: Stace (1997); Mitchell (1985)

occur. Churchyards and cemeteries are often similar in terms of wildlife, and both may attract wildlife because of their lower disturbance and greater habitat diversity compared to surrounding areas. For example, in Chicago, the number of bird species in cemeteries increased to a greater extent with increasing area of habitat than did neighbouring land (Lussenhop, 1977). In addition, birds nesting near to cemeteries with limited nesting sites, often foraged within the cemetery boundaries.

The gravestones themselves support a huge diversity of lichens, with over 500 of the 1,700 British species having been recorded from churchyards so far (Dennis, 1993). Vegetation overhanging gravestones in old abandoned areas may reduce lichen growth, as does moving or cleaning the stones. Older stones laid horizontally provide basking sites for reptiles including slow-worms (*Anguis fragilis*) and common lizards (*Lacerta vivipara*).

AQUATIC URBAN HABITATS

Many aquatic habitats in urban areas are similar to those in the wider countryside, although urban conditions often have major impacts on the organisms present. Habitats range from the artificial (e.g. sewage works), through analogues of more natural systems (e.g. sand and gravel pits, reservoirs and canals) to those natural habitats which run through, or were encapsulated by, urban developments (e.g. streams, rivers, ponds and lakes). One fundamental difference between such habitats is the rate of water flow. In lotic systems (i.e. running waters) flow rate is an important determinant of chemical status (e.g. oxygen and pollution levels) and therefore of plant and animal communities. Higher flows increase the rate at which oxygen diffuses into water and this promotes breakdown of organic pollution. Other pollutants may be quickly washed away. Lentic systems (i.e. still waters) are more influenced by sediment deposition and nutrient levels. Water bodies differ in terms of their nutrient levels. Oligotrophic water bodies are relatively nutrient-poor (especially in nitrogen) and are often characterised by a lack of organic substrate and few plankton, plants and animals. In contrast, eutrophic waters are relatively high in nutrients, have organic silty substrates, and high numbers of organisms. The effects of urbanisation – comprising run-off from roads (containing salt and oil), industrial and domestic effluents and impacts of stormwaters from areas of relatively impervious surfaces – combine to produce hostile habitats. Increases in population size and building density have raised discharges of both fresh and waste water into natural water courses. This impacts on more natural sites, increasing stream bed erosion and disturbing the aquatic habitat. If sediments or pollutants are associated with the discharge, then additional disruption occurs, and many aquatic environments fed from urban areas have poor water quality.

Many towns and cities have lost wetlands through drainage, and other aquatic habitats (e.g. reedbeds, marshes and coastal features) have been fragmented by urban development. Despite substantial improvements in recent years, many urban freshwater habitats are in a far worse condition than equivalent habitats in rural areas. However, there are still urban areas with water courses of considerable value, and those which do survive may be important components of urban habitat diversity, supporting a relatively rich flora and fauna (Plate 2.10). Habitats alongside aquatic environments are also important. Bankside vegetation and even patches of mud along the banks of lakes and rivers may feature specialist, and sometimes quite rare, plants (Chatters, 1996). Most of the plants in these transient sites are annual species which prefer open areas free of vigorous competitors. Often disturbance is essential to maintain these communities, which may contain species which were once widespread and are now quite rare, for example, small fleabane, mudwort, and pennyroyal.

Marshes and Reedbeds

Marshes and reedbeds occur where water does not drain away, where flooding occurs frequently, or where vegetation colonises the edges of aquatic habitats. In urban areas these habitats are often remnants of more extensive sites enclosed by development. However, due to an increasing awareness of the value to wildlife of reedbeds in particular, many are being encouraged at the edges of lakes, ponds and reservoirs. There are three important features that plants growing in these still waters need to be able to accommodate. These are anaerobic conditions (low oxygen concentration), nitrogen being only available in the form of ammonium (plants normally take up nitrogen as nitrates), and the presence of some

Plate 2.10: Urban canal with extensive fringing vegetation

chemicals in toxic concentrations. Some plants have physiological and biochemical adaptations to cope with these conditions. These include porous tissues with air spaces enabling water-covered roots and stems to obtain adequate oxygen, and the ability to immobilise toxins (such as iron) in their roots.

In waterlogged soils, mosses such as *Sphagnum* species grow alongside common cottongrass and cross-leaved heath. If the habitat features a lot of standing water, more aquatic species occur, ranging from floating species (e.g. common duckweed) and rooted plants (e.g. broadleaved pondweed), to emergent vegetation (e.g. water-plantain). Reedbeds are one of the more important edge environments of aquatic sites, providing habitats for aquatic and terrestrial organisms. Field voles, bank voles and shrews forage in amongst the reeds, whilst frogs (Species Box 2.22), toads and newts prefer open water. Of the invertebrates, freshwater shrimps and water hoglice are common detritivores. In more open water, dragonfly and damselfly adults fly overhead looking for prey, mates and egg-laying sites,

whilst the nymphs forage in the water below. Water beetles and aquatic bugs swim throughout the water column, caddis flies and several fly larvae live on muddy bottoms, and various snails feed on the vegetation. The dense structure of reedbeds provides shelter for birds including coots, moorhens, mallard and (on occasions) reed warblers, reed buntings and redshanks. The bittern, which is quite a rare bird in Britain, has been recorded around reedbeds in Wigan Flashes, an area resulting from subsidence of old coal workings (Smith, 1996).

Reedbeds have several properties that help to cleanse urban run-off (Brix, 1994). The cocktail of contaminants in run-off from roads, industry and stormwater includes suspended solids, nutrients (e.g. nitrates and phosphates) and toxins. As run-off passes through reedbeds, solids (including some particulate toxins such as heavy metals) are trapped by the mass of vegetation and incorporated into silt or adsorbed onto soil. Nutrients are often utilised by plants, decreasing the quantity that reaches open water.

Species Box 2.22: Common frog

Common frogs are 12–15 mm long just after metamorphosis, increasing to 20 mm by the end of the first year and 40 mm at the end of the next year. They are variable in colour (usually olive-green, grey or brown) with scattered dark patches and paler (white or yellow) undersides speckled with darker spots. This widespread species tends to live amongst damp vegetation except during the breeding season. They hibernate from October to February: males usually in mud at the bottom of ponds; females and juveniles often under stones, logs or piles of leaves. Migration to breeding ponds begins in February, with males usually arriving before females. Mating mainly takes place in March in shallow areas of ponds (usually the warmest places). The female lays a mass of eggs surrounded by jelly (about 1,000 eggs per mass) which are externally fertilised when the male secretes sperm over them. The egg mass usually floats, although it may become attached to vegetation. The jelly surrounding the eggs has channels within it allowing circulation of water round the eggs. After about two weeks, tadpoles hatch. At this stage, they have external gills which soon

disappear. During the following couple of months they metamorphose, growing hindlimbs, then forelimbs, and finally losing the tail. At this stage they leave the pond and move onto land. Following breeding, adults may stay in the pond for about a month, during which time they come onto land to feed. They are mainly active around dusk. They feed entirely on land on invertebrates, and any suitably sized moving prey is taken, leading to changes in diet composition with area and season.

Sources: Frazer (1983); Arnold and Burton (1978)

Reeds also absorb and assimilate some pollutants onto and into stem growth. This plant material dies back in winter, producing litter which is incorporated within the sediment. These cleansing properties of reedbeds are exploited throughout the world; both natural and constructed wetlands are used to control pollution levels in urban waters (e.g. Brix, 1994). However, unless silt is removed, reedbeds dry out. There may then be problems in disposing of contaminated silt.

Streams, Rivers and Canals

In clean running water, plants and animals are influenced by the rate of flow; fast waters scour substrate from the stream bed and reduce the material available for root growth and for animals to cling to or hide beneath. However, fast-flowing, turbulent water is rich in oxygen, which encourages algae and aquatic plants. Slower-moving water may have deep sediments providing a medium for rooting plants and burrowing animals. Very slow or stagnant waters, despite the lack of physical disturbance to organisms, often lack diversity as a result of low oxygen content.

Streams and Rivers

The majority of urban streams and rivers are in the lowlands where flow rates are relatively low. However, many are canalised over at least

Plate 1: Street trees

Plate 2: Formal layout of town park planting

Plate 3: Grey squirrel

Plate 4: Buddleja growing through a car park surface

part of their course, in some cases running in straight, artificial tunnels for much of their urban phase, causing an increase in flow rate. This is exacerbated by increased run-off from impervious surfaces (Marsalek *et al.*, 1993) and the lack of natural removal of rainfall by plant uptake (especially trees). Areas receiving urban stormwater have a reduced species diversity and fewer of those species typically associated with clean waters. These effects are not limited to events following heavy rain, since pollutants remain in the system during periods of low flow, often being more harmful since they are less dilute. In some cases, increased flow during stormwater events may in itself be more problematic than the pollutants, since the extra water removes plankton, increases turbidity (Seager and Abrahams, 1990) and causes erosion.

The plants growing along urban waterways increasingly include invasive species such as Japanese knotweed, Indian balsam and giant hogweed. Several other plants can be described as sewage species, the seeds of which remain viable after passing through the human gut. These are washed out with raw sewage during periods of high rainfall when combined stormwater and sewage drains overflow into rivers. Seeds of several such species (e.g. tomato, strawberry, fig and species of citrus) occur in silt from the banks of urban rivers such as the River Don in Sheffield (Gilbert, 1992c). Along a short section of the River Don, there are fig trees growing at the base of retaining walls. These trees are at least seventy years old and germinated when the river was warmed to about 20°C as a result of being used for cooling in the steel industry. Such plants are unlikely to germinate in the cooler waters found today. Other plants (such as apple and pear) found along urban waterways may also originate from sewage.

Rare species like water voles live along some urban rivers. In a study of urban stretches of the River Leen in Nottingham, Kirby (1995) recorded active burrows in over a third of the 13 km surveyed. This is surprising given that over half of the river examined was culverted or channelised, with banks which are unsuitable for burrowing. Since water voles have declined in Britain in recent years (Harris *et al.*, 1995), it is encouraging to find so many in an urban area. Mink also occur along stretches of some urban rivers and canals; this sometimes causes concern for other wildlife.

Canals

The major differences between canals and rivers are that canals have low flow rates and controlled water levels. The first true canals were built at the end of the eighteenth century to provide transport routes for freight, connecting urban areas with the coast, or with sources of fuel or other goods. For example the Bridgewater Canal was used to carry coal from Worsley into Manchester, and the Grand Union Canal links some of the larger cities in the Midlands and south of England (including Birmingham, Leicester and London). During the first half of the twentieth century, canals were more polluted than many rivers, due to low flow rates reducing dilution and aeration (Eaton, 1989).

Several introduced species have spread across the British canal system including Canadian waterweed from north America, water fern from tropical America, and invertebrates such as the zebra mussel and the amphipod *Corophium curvispinum*. Both of these invertebrates were originally estuarine species which now reside in many English canals (e.g. the Grand Union Canal). Many pond dwelling species have also colonised canals. One of these, the floating water-plantain, has spread from oligotrophic lakes to eutrophic lowland canals. Its success in canals is aided by management practices which retain open water and restrict the vigorous competitors which ordinarily thrive in eutrophic waters. The eutrophic waters of many lowland canals encourage plant growth, and open water is slowly colonised by floating and emergent vegetation becoming dominated by emergents such as reed sweet-grass. These communities are often lower in diversity and eventually silt up completely, unless kept open by traffic or management.

Boat traffic reduces the number of species present (Murphy and Eaton, 1983). Boats increase turbidity, reducing light available for photosynthesis both while material is suspended and when it settles on leaves. Abrasion by moving particles and direct impact by boats and their wash causes physical damage or uproots plants. Pollution from boats, including oil films, has less influence than other aspects of heavy boat traffic, but other pollutants (e.g. from industry) and shading by tall buildings and bankside vegetation are influential. Intensive management, which opens up the water to allow high levels of traffic, also reduces diversity, by removing colonising species even along the bankside (Briggs, 1996). Low level intervention keeps the channel clear, allowing moderate levels of navigation, whilst retaining nature conservation interest.

Several canals in urban areas have high nature conservation value and some, such as the Hollinwood Branch Canal in Greater Manchester and the Leeds and Liverpool Canal in West Yorkshire, are SSSIs (Eggo, 1990). The range of wildlife can be very impressive, sometimes including significant proportions of the national population of rare species such as floating water-plantain and tufted loosestrife (Briggs, 1996). Both aquatic invertebrates and fish are vulnerable to pollu-

tion. Fish present include native and introduced species such as the predacious pike-perch which was illegally introduced into canals in the Midlands during the 1970s. Water voles live along even heavily used and quite polluted canals, whereas water shrews are restricted to cleaner waters. Mink are possibly as common in canals as they are in rivers. Bankside habitats add to the diversity in rural canals, but are often less evident in urban areas. Built structures (e.g. bridges, tunnels, aqueducts, etc.) provide roosting sites for a variety of bats.

Ponds, Lakes and Reservoirs

Although natural lakes are infrequent in urban areas, open water habitats in the form of flooded gravel pits, reservoirs and ponds are common in towns and cities and provide similar conditions to natural still waters (Plate 2.11). Since larger ponds and lakes are relatively closed environments, they tend to accumulate nutrients and become eutrophic. In such nutrient-rich waters algae thrive, sometimes leading to large populations which may become dense enough to see (algal blooms). Increased algal and plant growth reduces the available light and oxygen levels. Some invertebrates are intolerant of these conditions, as are many fish, consequently wildlife in nutrient-rich sites is very different to that in oligotrophic waters. Some algae produce toxins, which are potent enough to kill fish and even domestic animals. While not normally fatal to humans, they can cause serious illness and are unpredictable in timing and toxicity.

Much open water, especially in relatively small and shallow sites, is vulnerable to development. In Portland, Oregon, out of 233 small wetlands identified in 1981/2, only 141

Plate 2.11: Fishing pond near to inner city housing estate

remained ten years later (Holland *et al.*, 1995). Most of the losses were due to human activities (mainly development of urban land). A quarter of the wetlands still present had been seriously degraded, and many others were disturbed. Similar situations occur in Britain: Leicester lost 78 per cent of its ponds over the last century (Owen, 1991), although some new ponds have been created in recent years.

Ponds

Ponds are water bodies up to about 2 ha in area that hold water for at least four months each year. Despite their small size, they are valuable wildlife resources especially in urban areas. A species list drawn up for the small (20 × 20 m and 1 m deep) artificial pond at William Curtis Ecology Park in London before the site was redeveloped identified a large range of animals (Table 2.8). Urban ponds provide homes for a range of amphibians, many of which have been adversely affected by habitat destruction elsewhere. Common frogs, common toads and several species of newt live in small garden ponds. The commonest newts in small ponds are smooth newts, and to a lesser extent palmate newts, although several introduced species (especially alpine newts) live in the south. Small ponds are less suitable for great-crested newts (Species Box 2.23) which tend to live in larger ponds, reservoirs and lakes. All British newts, and especially great-crested newts, have declined during the last fifty years, and all are protected under the Wildlife and Countryside Act (Department of the Environment, 1981). The great-crested newt is fully protected under the act, including prevention of disturbance and destruction of its habitat.

A familiar bird on open water is the mallard (Species Box 2.24). Urban mallards have adapted to human impacts, and indeed are often encouraged by artificial feeding, most of which occurs during the autumn and winter. Compared to rural populations, urban mallards in New Jersey are heavier, begin nesting earlier and continue later (utilising both natural and artificial structures) and are more tolerant of disturbance (Figley and VanDruff, 1982). They generally have a higher nesting success because there are fewer egg-eating predators, although in some areas brood mortality is higher, for example in public parks, because of predation by cats as well as disturbance by visitors and pets.

Table 2.8:   Pond life in the William Curtis Ecological Park pond

Taxon	Comments
Flatworms	Several species were present from 1977.
Molluscs	At least seven species, mainly gastropods, but also some bivalves (*Pisidium* species) were present. In some cases large numbers were found, although these fluctuated with time (e.g. *Lymnaea stagnalis* and *Planorbis planorbis*). Mainly introduced, either deliberately or on plants.
Leeches	Three species found: *Erpobdella octoculata* was the most frequent and probably colonised naturally from the Thames.
Crustaceans	Several species were found in large numbers, most having been introduced. Water hoglice and freshwater shrimps were the most common larger species, although they fluctuated in numbers over time, becoming quite scarce on occasions. Water fleas were much more successful.
Insects	Eight dragonflies and three damselflies were recorded: the common darter and blue-tailed damselfly being the most successful breeding species. Mayfly larvae were recorded in the first year. Common species of water boatmen (*Corixa punctata* and *Sigara dorsalis*), backswimmers (*Notonecta glauca*) and pond skaters (*Gerris lacustris*) were present from the early stages. Several fly larvae including rat-tailed maggots and chironomids were recorded. Occasional caddis flies were found. Several water beetles have been recorded, together with some ground beetles around the edges of the pond.
Arachnids	Lycosids were recorded amongst surface vegetation. Mites were particularly common in the summer.
Fish	Three-spined sticklebacks were introduced in the first year, but were replaced by the nine-spined sticklebacks introduced about a year later.
Amphibians	Both palmate and smooth newts were introduced early on. Neither established at first, possibly due to predation by sticklebacks. Smooth newts were the more successful eventually. Both common toads and common frogs were successfully introduced and became common breeding species.
Birds	Both moorhens and mallards bred on or near the pond. Other species fed along the edge or used the water for bathing.

Note: Records began in 1977 when the pond was built and stopped in 1985 when the site was redeveloped.

Source: Emery (1986)

Reservoirs

Although few reservoirs lie directly within urban areas, there are many on the urban fringe and many which are under recreational pressure. In a survey of the 537 larger (over 2 ha) reservoirs in England and Wales in 1976, 344 had some kind of recreational pressure, mainly angling (327), although other activities such as sailing, SCUBA diving and canoeing took place at 116 (Hengeveld and de Vocht, 1980/81). Activities such as sailing may adversely affect waterbirds, causing some species to decline (e.g. mallard) and others (such as teal and wigeon) to avoid the area when sailing activities intensify (Edwards and Howell, 1989). However, winter use by mallards, pochard and tufted ducks and breeding by great-crested grebes are maintained if there

Species Box 2.23: Great-crested newt

The great-crested or warty newt is the largest British newt (up to 140 mm in length). They are dark above with yellow-orange undersides marked with black patches. Breeding males have a jagged crest along the back and tail, with a break at the base of the tail. Although they can be found in water throughout the year (preferring still or slow-flowing water with good reed growth), after July they more often inhabit damp areas under logs and stones where they hibernate over winter. They return to the water around March to breed, with males tending to arrive first. Unlike frogs and toads, fertilisation is internal; the male deposits sperm in a bundle (spermatophore) which he encourages the female to pick up by guiding her over it until her cloaca (reproductive opening) makes contact with it. Once inside the female, the spermatophore breaks down and releases sperm to fertilise the eggs, which are then laid one at a time amongst aquatic vegetation. Newly hatched tadpoles possess balancing organs which point forward from the sides of the head and help them to attach to vegetation. After a few days, swimming ability improves and these organs are lost. Younger tadpoles tend to keep to deeper water. Metamorphosis takes place over the next two to three months with forelimbs appearing first, and hindlimbs after about seven weeks. Unlike frogs and toads, the tail is retained. If eggs are laid late in the season, the young newts may overwinter in the water, completing metamorphosis the following spring. Under normal circumstances the

young leave the water at the end of the summer, and do not return until they are sexually mature, usually two (and up to four) years after metamorphosising. Tadpoles are carnivorous, feeding at first on small aquatic organisms (e.g. fly larvae, ostracods and copepods) and later on larger animals including the young of other amphibians (especially frogs). Adults in ponds feed on similar (but larger) prey items to the young. When on land they consume a variety of invertebrates.

Sources: MacGregor (1995); Frazer (1983)

are large areas of shallows inaccessible to boats and especially where these are screened from, and less disturbed by, boating areas. It is even more important to have undisturbed areas if shore fishing also takes place. Wigeon avoid the shore line where anglers are present and, if sailing precludes them from the centre of the water, they may move off the site completely. Reservoirs with steep, reinforced-concrete banks provide unsuitable

conditions for many emergent or rooted plants, waterfowl, or those fish which require shallow, vegetated areas for spawning (Kubečke and Duncan, 1994). The best mixture of depths provides shallows (less than 0.35 m) for dabbling ducks (e.g. teal) and deeper water (up to 2.5 m) for diving ducks (e.g. tufted ducks). Other birds which have increased on reservoirs in recent years are black-headed and common gulls.

··

Species Box 2.24: Mallard

The mallard is one of the commonest birds of urban waters and is Britain's most urban duck. It is around 580 mm in length, with a mass of around 1.1 kg. The male has a yellow bill, a shiny green head, white collar, purplish-brown breast, with a grey back and underparts. The female is less colourful, having speckled brown plumage and a brownish bill (with some orange on the sides of the bill). Both sexes have a whitish tail, orange legs and a blue band on the upperside of the hind edge of the wing (most obvious in flight). They are most commonly found on still, shallow waters. Mallards breed in their first or second year from March to June, often producing a second brood if the first fails. Between nine and thirteen eggs are laid per clutch; urban clutches may be at the top end of this range. The female builds the nest often in ground cover but sometimes in the open or in trees. She does most of incubation (twenty-four to twenty-eight days) and subsequently looks after the young for around eight weeks, after which they are able to fly. Mallards undergo an annual flightless moult lasting about thirty-three days during which birds stay on or near water, or hide to avoid predators. They are highly omnivorous, feeding mainly on seeds and vegetation with some animals (including Crustacea, insects and molluscs). Feeding takes place either in water (under 1 m deep) or on land where they graze leaves and shoots.

Sources: Figley and VanDruff (1982); Cramp (1977)

··

Sand and Gravel Pits

The demand for sand and gravel for industry and the building trade has led to about 15,000 ha of abandoned (now flooded) pits, the total area of which increases at about 500 ha per year (Andrews, 1991). These flooded pits are important for overwintering wildfowl such as

Table 2.9:   Wildfowl using gravel pits in Britain

Species	Approximate wintering numbers	Feeding and diet
Mallard	60,500	
Teal	5,800	Omnivorous, surface-feeding, dabbling ducks
Gadwall	540	
Wigeon	3,400	Herbivorous, surface-feeding
Shoveler	1,060	Carnivorous, surface-feeding
Tufted duck	15,120	
Pochard	14,600	Omnivorous (mainly invertebrates), diving ducks
Goldeneye	630	
Goosander	540	Carnivorous (mainly fish), diving
Mute swan	2,900	Herbivorous grazer on land and in water
Canada goose	9,080	Herbivorous grazer on land

Source: Street (1985)

tufted ducks and pochard (see Table 2.9). Little ringed plovers were first recorded breeding in Britain in 1938 and have since colonised old gravel workings throughout England and Wales which now account for 60 per cent of the breeding population. Great-crested grebes were previously thought to be threatened in Britain but are now increasing, and 30 per cent depend on gravel pits. Other birds found include common terns which use them as inland nesting sites. Many other wetland species benefit from these sites and some pits are designated as SSSIs. Aquatic and emergent vegetation attracting dragonflies and other aquatic invertebrates, together with the plant and animal communities of the banks and more terrestrial habitats beyond, often creates rich and diverse habitats.

Sewage Works

Sewage works are perhaps one of the last places normally associated with wildlife value. However, a wide range of invertebrates live in the large tanks where sedimentation of solids occurs, and especially where shallow lagoons are used for final treatments. Segmented worms and the larvae of various midges feed on decaying organic matter in the tanks while adult midges and other flies are found in the air above, looking for mates or egg-laying sites. These animals, together with decaying organic material and algae growing on it, provide a rich source of food for birds. Starlings are one of the commonest species and may comprise over 75 per cent of the birds present (Fuller and Glue, 1981). Together with corvids (usually magpies and carrion crows) and wagtails (mainly pied, but also grey and yellow), they feed among the pebbles in filter beds

Plate 2.12: Sewage farm filter bed

(Plate 2.12). Swifts, martins (sand and house) and swallows feed on aerial insects above the tanks and lagoons. If larger effluent lagoons are present, several waterbirds such as ducks (teal, mallards, shoveler and tufted ducks) may occur. However, the lack of open water in modern sewage works reduces the diversity of birds present (Fuller and Glue, 1981).

The fertile nature of the open land within sewage works makes it ideal for nutrient-loving species such as common nettles (Species Box 2.25). Nettles are rich in insects and other invertebrates. Around thirty species of insect are more or less restricted to common nettles, with over seventy-five more being commonly associated with the plant (Davis, 1991). Nettles are a major food resource for the larvae of many moths (such as the snout, burnished brass, silver Y and Hebrew character). It is the food plant for four British butterflies (the small tortoiseshell, peacock, red admiral and comma), which have shown a slight increase in abundance recently whilst many related non-nettle feeders have declined. Bugs and snails also feed on nettles, as do some beetles such as

weevils. Predators include other beetles (rove beetles and seven-spot and two-spot ladybirds) together with harvestmen and spiders.

Coastal Features

Many urban coastal areas are under pressure from human impacts such as pollution, dredging, marinas, industry, coastal defences and other coastal engineering. Most of the twenty-seven important marine areas around England are affected by fishing and recreation, many have sewage outflows, and some are influenced by industrial effluents or mineral exploitation (Marine Task Force, 1994). Consequently, large areas of coastal habitat are threatened (Figure 2.7). Over 50 per cent of south-east England's coastline is protected by sea walls and other artificial structures (Doody, 1992) which directly affect habitats such as saltmarsh, where the high tide level effectively becomes the position of the wall. This, along with sea level rise (due to sinking land levels and global warming), reduces the

Species Box 2.25: Common nettle

The common (or stinging) nettle is a perennial species growing up to 1.5 m tall. It is a widespread native plant, and is a good competitor, found especially in woodland and fertile sites, including cultivated ground and around animal latrines. Although often associated with relatively undisturbed areas, nettles are also strong colonisers of heavily disturbed fertile areas. It is mainly found in damp areas, but may be present in wetter shaded habitats (e.g. the fens). It is particularly associated with sites rich in phosphates. The nettle is dioecious (has separate male and female plants) and forms green, inconspicuous flowers which appear in June and July, and sheds seed from August. The leaves are heart-shaped and toothed, arranged opposite one another. They contain high levels of nitrogen, calcium, magnesium and iron. Both leaves and stems have stinging hairs which protect against some vertebrate grazing, but not from invertebrate attack. It is an important plant for many insect species, providing food for herbivores (especially butterflies and moths), and prey and shelter for predators. Nettles also reproduce vegetatively from rhizomes and often

form large clumps. Because of the density of both the living canopy and the fairly persistent stem litter, it inhibits other species, and is often found in monoculture stands.

Sources: Stace (1997); Corke (1991); Grime *et al.* (1988)

amount of mudflats available to feeding birds. There are also indirect effects of coastal protection; the reduction in erosion of one area of coastline reduces the material deposited elsewhere. Dynamic coastal processes are then interrupted and natural habitats are lost. High levels of pollutants enter coastal areas, some via estuaries from industries occupying adjacent land (often chemical and petrochemical companies), and others through outflows into bays, dumping at sea (e.g. sewage) and accidental spillages at sea (e.g. oil). These adversely affect coastal wildlife and, in the case of oil spills, can be severe and long lasting. Even operations to clean up after human impacts can themselves cause problems. The removal of litter using mechanical beach-cleaners, also removes naturally occurring strand-line material including pioneer plant species (Llewellyn and Shackley, 1996). Invertebrates also live in these areas, under and on organic material. In some areas these include relatively rare species (e.g. the woodlouse *Armadillidium album*, and the ground beetles *Broscus cephalotes* and *Nebria complanata*).

Docklands

Many of Britain's urban areas have a substantial amount of disused dockland: in Liverpool, the South Docks stretch for about 2 km along the River Mersey. The water in docks is saline at the coast, brackish when fed by tidal

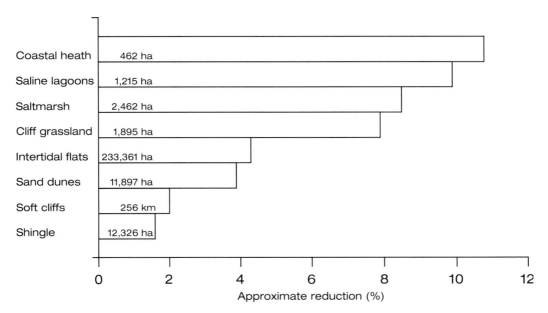

Figure 2.7: Anticipated reductions in the extent of coastal habitats by 2013. Numbers in bars indicate the approximate extent in 1993.
Source: Pye & French (1993)

estuarine waters and freshwater further inland. In many cases, the conditions are harsh, with large fluctuations in temperature and oxygen content, and often with poor water quality. Many docks have been re-developed, and the water quality improved by dredging and removal of contaminated sediments, and control of water interchange with polluted rivers or estuaries. This management reduces periodic oxygen depletion and algal bloom development. The resulting sites are lagoon-like habitats with great potential for conservation and education. The fauna and flora colonising the walls and sediment of two docks in Liverpool increased in diversity and cover during the six years following restoration (Allen *et al.*, 1995). The total area (nearly 30 ha) contained over ninety species, including some lagoonal specialists (e.g. the bryozoan *Conopeum seurati*, the amphipod *Corophium insidiosum*, and the prawn *Palaemonetes var-*

ians). Docks are also important for water-birds: from December to February, 1,100 to 1,400 pochard per month visit Salford Quays (alongside the Manchester Ship Canal) along with tufted ducks, breeding mute swans, and occasional common gulls, scaup, shelducks, ruddy ducks and oystercatchers (Smith, 1996). However, the depth and topography of typical docks (i.e. lacking shelving sides and adjacent grassland) makes them less suitable for geese, waders and many species of ducks.

A surprising resident of English docklands is *Euscorpius flavicaudis*, the only scorpion found in Britain. It is common in western Europe, and has been introduced into docks in south-east England. In Sheerness Docks (Isle of Sheppey in Kent), *E. flavicaudis* has persisted for many years, and now has a population of 600–700 animals (Benton, 1992). The scorpion resides in deep crevices in walls with one individual per crack. They shelter from adverse conditions deep within the crack, and wait at the entrance for passing prey of woodlice, spiders, insects and even other scorpions. Although it is a typical scorpion, some 35–45 mm in length, it is not harmful to humans.

Estuaries

British estuaries are highly important habitats, comprising about 28 per cent of the estuarine area of the European Atlantic and North Sea region (more than any other European country). In England 30 per cent of SSSIs contain estuarine habitats (Vincent, 1992). They are particularly important for waterbirds in winter, especially from December to February, when over 1 million waders (90 per cent of the British population) and 580,000 wildfowl (38 per cent of the British population) use British estuaries as feeding grounds. The main concentrations of estuaries are along the North Sea coast, the Solent, the Severn and in north-west England. Estuaries to the west and on central and northern North Sea coasts tend to have sandy habitats. Those along the Channel and southern North Sea coasts have more mudflats (i.e. areas of generally bare sediment between the mean high and low water marks). Human pressure on estuaries is considerable and increasing. Encroachment for development, including land claim, is one of the most important threats. Urban and industrial development of docks, barrages, power generation, mineral extraction, petrochemical industries, pollution and recreation all affect the estuarine environment. Land claim and accretion have combined to reduce the area of some estuaries: the docks on the Dee Estuary at Chester, Shotwick, Burton and Parkgate are now abandoned due to silting up (Doody, 1992).

Estuarine mud contains high densities of invertebrates which provide food for birds: per square metre of sediment there can be 36,000 laver spire shells, 22,000 *Corophium volutator* (an amphipod) and 220 lugworms (Andrews, 1989). Different invertebrate species live at specific depths and are fed on by different birds. Lugworms are taken by long-billed species such as curlew, birds such as dunlin with intermediate bills feed on *C. volutator*, and plovers (which are short-billed) feed on surface species such as laver spire shells. Estuarine plant communities are dependant on the tidal range; those above the high tide line (only infrequently inundated by the sea) support the richest communities. In time these areas develop into entirely terrestrial environments. Succession of communities depends on the distance from the water line and the stage of development. Early colonists include glassworts, followed by common saltmarsh-grass and sea aster, then common sea-lavender, thrift and sea-purslane.

Some areas in north-eastern England have lost over 90 per cent of their intertidal land over the last 100 years. Some birds have especially suffered from losses of estuarine habitats; two-thirds of British redshanks (30,000 pairs) breed on saltmarshes. Reductions in tidal range (e.g. resulting from the building of barrages across estuaries) reduce the amount of intertidal substrate available for winter feeding sites for many shorebirds (Ferns *et al.*, 1984). Barrages are intended to provide protection against flooding (e.g. the River Thames), the potential for power generation (e.g. the Severn Estuary) and areas for recreation and commerce (e.g. the River Tees). However, their development influences the organisms living upstream. Pollutants, both inorganic (e.g. metals) and organic (e.g. hydrocarbons), may accumulate in areas of restricted river flow such as impounded areas of the River Lagan in Belfast (Watts and Smith,

1994). However, pollutant concentrations may not increase, at least partially due to dilution by the increased volume of water behind the barrage (Gray, 1992). The combination of pollution from upstream and developments along the sides of estuaries can concentrate some toxins. Such accumulations may reduce the food available for wildfowl, although organic pollutants such as sewage can encourage some invertebrates, providing an abundant source of food for waterbirds. Similarly, reductions in water turbidity could encourage phytoplankton and hence some invertebrates which are prey for some bird species (Gray, 1992). The spread of common cord-grass (a vigorous hybrid between small cord-grass and smooth cord-grass) in estuarine areas is a cause for concern. The hybrid, which was originally planted to aid reclamation, colonises bare mud thus reducing the feeding areas for many birds. Control is difficult, since although herbicide treatment is possible, it is expensive over large areas.

Other Coastal Habitats

Saltmarshes are typified by halophytic vegetation, the species composition of which changes from the mud or sand flats at the seaward side to terrestrial habitats or artificial embankments landward. One remnant area of salt marsh in Gateshead, where the River Team flows into the River Tyne, is flooded monthly by the tide. In places where salt water is the main influence, sea aster and sea club-rush grow, and where freshwater enters, there is a reedbed which is home to many invertebrates, upon which redshanks feed. Many saltmarshes are lost each year, mainly due to land claim for agriculture, housing, marinas and industry (Pye and French, 1993). Land reclaimed in the past may now have some conservation value. For example, coastal grazing marshes are flat, low-lying pastures reclaimed from tidal saltmarsh using sea walls and drained using ditches. These often flood in winter, and support plants such as least lettuce, divided sedge, soft hornwort, and animals like scarce emerald dragonflies, teal and redshanks. In London, 65 per cent of such coastal grazing marshes have been lost over the last 50 years (8,700 ha) with urban development accounting for half of the loss. One of the major potential effects of proposed estuarine barrages is erosion of salt marshes and mud flats (Gray, 1992).

Sand dunes are important coastal habitats vulnerable to urban impacts. There is a gradation from the strand-line landwards of mobile sand dunes, semi-fixed dunes, fixed dune grassland communities, fixed dune sand sedge communities and dune slack communities. The major losses of these habitats are to waste dumping and residential and recreational developments (Pye and French, 1993). Sand dunes are prone to erosion and many are damaged by walkers and motorbikes with uncontrolled access. Erosion may be controlled using fences to trap the sand, then planting with marram (which is tolerant of dry conditions, i.e. is xerophytic) and has long roots which help to bind the sand, so stabilising the dune. Many soft cliffs are now protected by sea defences to guard developments from erosion. The cliff tops are often grasslands or heathlands (containing heather and bell heather on drier soils, and in the north these together with cross-leaved heath or crowberry). Many cliff grasslands and coastal heaths are threatened by developments including car parks, caravan sites and agricultural expansion (Pye and French, 1993).

Shingle shores mainly comprise pebbles, with a mix of sand and cobbles (defined as having a mean grain size of 2–200 mm), and have features such as spits, barrier beaches and forelands. These features are under threat by aggregate extraction and development for housing and recreation. In natural systems, shingle is replaced by sediment eroded from soft cliffs. Sea wall construction reduces the supply of sediments, and walls at the top of beaches increase wave action on the foreshore, which removes shingle (Pye and French, 1993). One example where human involvement interferes with the natural movement of shingle is the shingle shore west of the nuclear power stations at Dungeness in Kent. To protect the complex from erosion, shingle is relocated from the east to the west while erosion moves it back again (Doody, 1992). The shingle at Dungeness is more extensive than in many

areas of coastline and forms vegetated ridges which show a succession of plants (Morris and Parsons, 1993). The vegetation ranges from specialists (e.g. sea kale and spear-leaved orache) on younger nutrient-poor ridges, through false oat-grass swards, to shrubby communities containing broom (a nitrogen-fixing species). The hostile conditions (dryness and temperature extremes) on shingle encourages unique communities of plants and invertebrates (sixteen of the nineteen species of British bumblebees and nineteen species of ants occur at Dungeness). The removal of shingle in recent years has destroyed over 40 per cent of these ridges and disturbed more of the vegetation.

3

MANAGEMENT AND CONSERVATION

•

Urban habitats often appear beset by problems. Pollution, the spread of noxious species, dereliction and the need for reclamation, restoration, habitat protection and species conservation are all important in urban environments. However, there are also many opportunities for wildlife in novel artificial habitats or by the creation of new habitats. In addition, there are benefits in integrating human communities within urban habitats, not least for environmental education. The following sections discuss several of these issues with reference to example habitats and species within urban areas.

POLLUTION

Pollution is the occurrence in the environment of by-products of human activity which may adversely effect health or a desirable habitat, species or community (e.g. Plate 3.1). In urban areas the environmental consequences of pollution can be severe. The three examples discussed below examine the impacts on wildlife of different types of urban pollution.

Heavy Metals and Woodlands

North-west of Bristol, at the point at which the River Avon joins the River Severn, is the large industrial area of Avonmouth. Amongst the industries is a large smelting works which

continues a history of some seventy years of metal smelting in the region. Following an expansion of the works in the late 1960s, environmental concerns lead to investigation of the environmental impact of the smelter. In 1984, annual production capacity was around 90,000 tonnes of zinc, 40,000 tonnes of lead, 360 tonnes of cadmium and 162,000 tonnes of sulphuric acid (Read, 1987). Stack emissions of $6.0 \, \text{kg h}^{-1}$ zinc, $4.0 \, \text{kg h}^{-1}$ lead and $0.4 \, \text{kg h}^{-1}$ cadmium combined with pollution from dusts derived from the raw materials and impacts from neighbouring sources, including a refuse incinerator. The local area therefore received large amounts of atmospheric pollutants, mainly in particulate form. Many of these were heavy metals (including copper, cadmium, lead and zinc) which, although often necessary in very small quantities for metabolic processes, are toxic to many species.

The impacts of four pollutants (cadmium, zinc, lead and copper) were examined in woodlands north-east of the smelter (the direction of the prevailing wind). Lead tends to remain associated with surface soil organic layers, whilst zinc and cadmium are more mobile, increasing the probability that organisms take them up (Martin and Coughtrey, 1987). Susceptible species tend to be absent from contaminated woods and even tolerant animals are only found at low densities. The low densities of soil or surface-active species which feed on leaf litter and other organic debris mean that leaf litter layers are not quickly

Plate 3.1: Polluted urban canal

consumed and become quite deep in contaminated woodlands. *Nebria brevicollis* (Species Box 3.1) is a common ground beetle which preys on invertebrates in leaf litter, and normally avoids low levels of prey during dry summer conditions by remaining inactive (summer diapause). The diapause does not occur in contaminated woods, possibly because the beetles cannot catch sufficient prey prior to diapause to provide the fat stores necessary to survive a dormant period (Read *et al.*, 1987).

In some polluted habitats, bioaccumulation of toxins occurs between trophic levels; herbivores and decomposers take up pollutants from vegetation, and predators accumulate it from their prey, each stage having successively higher levels. A tendency to accumulate pollutants makes some species good indicators: some woodlice (especially *Porcellio scaber*), which are decomposers of dead leaves, accumulate zinc, cadmium, lead and copper within part of their body tissue, the hepatopancreas (Hopkin *et al.*, 1986). In six woodlands at different distances from the

Avonmouth smelter there are indications that decomposers are less strongly influenced by heavy metal contamination than are some predators (Read *et al.*, 1998). Contaminated sites are higher in all metals at all trophic levels than are clean sites further from the source of pollution (Figure 3.1). Although accumulation occurs in some soil layers and some decomposers, the concentration does not increase throughout the trophic levels. Shrews (which are insectivores) from contaminated sites contain large amounts of lead and cadmium, but not much higher than some of the invertebrates upon which they feed (Read, 1987). Different physiological (e.g. storage in appropriate tissues and excretion) and behavioural responses (e.g. avoidance) may complicate the picture. Centipedes fed on contaminated woodlice accumulate cadmium to lethal levels, although *Dysdera crocota*, a spider which is a specialist predator of woodlice, does not assimilate heavy metals from contaminated prey, probably by avoiding those parts rich in pollutants (Hopkin and Martin, 1985). The effects of contamination,

· ·

Species Box 3.1: *Nebria brevicollis*

Nebria brevicollis is a very common, medium sized (10–14 mm) ground beetle (of the Carabidae family). It is black with a short pronotum (upper surface of the thorax). Its usual habitat is deciduous leaf litter, but it is wide ranging and often present in gardens, especially in damp areas. Adults are active from February until December with activity peaks from April to June and August to September. They may aggregate, especially under logs and stones. There is usually a diapause period during the end of July and August, possibly to avoid periods of low food availability (but see text). Breeding usually occurs in autumn, with larvae overwintering and emerging as adults the following spring. However, spring breeding with autumn adult emergence has been noted in wet habitats. The larvae have very long cerci (appendages attached to the hind end of the animal) with long hairs and powerful legs (especially the hind pair). They tend to prefer damp, shady places where they burrow into the soil. *N. brevicollis*

is entirely predatory, feeding on small invertebrates such as springtails and other soil animals including worms.

Sources: Forsythe (1987); Thiele (1977)

although complex, can be severe, with some species being absent from heavily polluted sites (Table 3.1). The smelter recently reduced the output of these pollutants and some areas are now less contaminated than before. This may have affected the wildlife, and wolf spiders, found only in low numbers in polluted woodlands, have recently been recorded from open ground near to the smelter.

Lichens and Sulphur Dioxide Pollution

Sulphur dioxide (SO_2) is a major urban pollutant, occurring in vehicle exhausts and both industrial and domestic discharges. Despite a decline in levels from the 1950s, by 1988 UK sulphur deposition was still one of the highest in western Europe, with up to 88 per cent coming from domestic sources (United Nations Environment Programme, 1991). Levels have remained fairly constant since the mid-1980s, although recent figures for 1993/4 indicate that many cities still have high concentrations especially near roads, where levels often exceed World Health Organisation guidelines (Bower *et al.*, 1995). It has been known since the early part of the twentieth century that lichens are particularly sensitive to sulphur dioxide pollution, and that lichen cover decreases with proximity to urban centres. This sensitivity is due, in part, to the ease with which lichens absorb atmospheric pollutants, because they lack cuticles (unlike plants) and take up nutrients directly from the air.

In extremely polluted areas, lichens do not grow on trees (especially trees with moderately acidic bark such as oaks). Instead, a bright green alga (*Desmococcus olivaceus*) covers the base of the tree. As pollutant levels fall, lichens

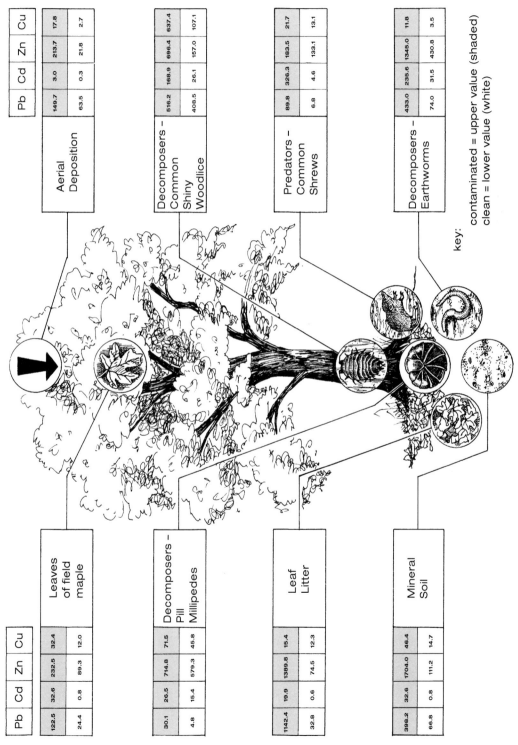

Figure 3.1: Levels of heavy-metal contamination in components of two woodlands.

Note: All values are in µg g^{-1} dry weight

Source: Read (1987)

Table 3.1:   Woodland invertebrates and heavy metal contamination

Taxon	Species common in contaminated woods	Species common in less contaminated woods
Millipedes	*Chordeuma proximum*	*Tachypodoiulus niger*
	Glomeris marginata	*Ophyiulus pilosus*
Harvestmen	*Leiobunum blackwalli*	*Mitosoma chrysomelas*
	Leiobunum rotundum	*Analasmocephalus cambridgii*
	Nemastoma bimaculatum	
Spiders	*Coelotes atropos*	*Pisuara mirabilis*
	Robertus lividus	*Pardosa nigriceps*
	Lepthyphantes zimmermanni	*Pardosa pullata*
Ground beetles	*Cychrus caraboides*	*Carabus granulatus*
	Carabus nemoralis	*Bembidion tetracolon*
	Carabus violaceus	*Pterostichus nigrita*
	Agonum albipes	

Source: Read (1987)

first appear low on tree trunks (where there is most shelter from pollution) and then higher up (above 0.5 m from the ground), and eventually on twigs (the most exposed part of the tree). Crustose species (see Species Box 3.2) are the most tolerant of polluted areas because they grow close to the surface and therefore have a relatively small surface area exposed to the pollutant. In less polluted areas, foliose species may survive, while fruticose species are only found in clean areas. These latter two types have large surface areas exposed to atmospheric contaminants. These differences in tolerance are used to monitor pollutant concentrations, and zones have been identified based on the presence of specific lichens in relation to sulphur dioxide levels (Richardson, 1992). The most comprehensive and widely employed zone scale for England and Wales is summarised in Table 3.2. This scale relates to deciduous trees with rough bark of moderate acidity, such as oaks and ashes, situated in fairly open areas. Where the bark is nutrient-rich (in elms, London plane, beech, sweet chestnut or horse chestnut) or where bark is impregnated with agricultural or other nutrient dusts, then an alternative scale is needed (see Hawksworth and Rose, 1976). Under these circumstances, lichens such as *Xanthoria* (which only occur in nutrient-rich environments) may be present.

The pollutant related changes in lichens affects species which associate with them. A decline in the snails which feed on lichens (such as *Balea perversa* and *Pyramidula rupestris*) has been recorded in areas of high sulphur dioxide pollution (Willing, 1993). Reduction in lichen cover and diversity has also been implicated in the distribution of different colour morphs of the peppered moth (see earlier discussion). As sulphur dioxide levels fall, less tolerant species are recolonising urban sites. However, the patterns of recolonisation do not always follow the pattern suggested by the level of tolerance (Richardson, 1992) because the dispersal capabilities of the species

··

Species Box 3.2: Lichens

Lichens are symbiotic organisms, part fungus and part single-celled algae (phyla Chlorophyta) or blue-green algae (Cyanobacteria). There are many more fungal species involved in lichen partnerships than algae. The majority of algal partners tend to be *Trebouxia* (or similar species of chlorophytes) or Cyanobacteria such as species of *Nostoc*. The fungus forms the outside layer and provides the alga with protection from extreme conditions. The alga produces sugars by photosynthesis and provides the fungus with nutrients. In almost all cases, the fungi (and some of the common algae) associated with lichens are never found alone. Although sexual reproduction occurs in some lichens, vegetative reproduction is more common. In most cases vegetative

reproduction incorporates both fungal and algal cells, avoiding the necessity for the fungus to locate new algal cells and dispersal is often by wind-blown fragments. Lichens can be divided according to their growth form: crustose lichens form crusts which adhere tightly to the substrate and cannot be removed except by scraping; foliose lichens form leafy clumps which sit proud of the substrate and may be detached using a knife; and fruticose lichens form bushy or shrubby outgrowths which sit proud enough to be picked off by hand. Lichens occur across Britain and can be found on soil, rocks (including walls, gravestones and buildings), bark and wood.

Sources: Dobson (1992); Richardson (1992); Hawksworth and Rose (1976)

···

may restrict their reappearance in towns and cities.

Freshwaters and Pollution

In the first half of the twentieth century urban freshwaters were highly polluted. Since then, pollution from domestic and industrial sources has been reduced, although eutrophication and acidification have increased (Eaton, 1989). Eutrophication has increased with re-

lease of treated sewage and urban sources of nitrate and phosphate. Hengeveld and de Vocht (1980/81) describe six groups of water-borne pollutants in urban systems (see Table 3.3), ranging from those with slight toxicity and ecological affects to artificial pollutants with strong persistence and mobility.

Such pollutants may severely affect the wildlife present, since many aquatic species including invertebrates (Table 3.4) and fish (Table 3.5) are intolerant of high pollution levels. The type of pollutant, the amount discharged,

Table 3.2:   Lichen zone scale for mean winter sulphur dioxide estimation

Zones		Characteristic lichen species present	Mean winter SO$_2$ levels ($\mu g\ m^{-3}$)
Lichen 'desert' (0–1)	0	No epiphytes present on trees	Very high
	1	Tree bare except for green algae (*Desmococcus olivaceus*) at base (lowest 0.5 m) of tree	Over 170
Inner transition zone (2–3)		Green algae found higher up the tree. Encrusting lichens found at base of tree, each found higher up the tree as others move in.	
	2	Only *Lecanora conizaeoides* is found	About 150
	3	*L. conizaeoides* and *Lepraria incana* are found	About 125
Outer transition zone (4–5)		Green algae and encrusting lichens found higher up the tree. Foliose lichens at base of tree, each found higher as others move in.	
	4	*Hypogymnia physodes* and/or *Parmelia saxatilis* or *P. sulcata* plus others found only at base of tree	About 70
	5	The species above and others appear higher up the tree and *Parmelia glabratula, P. subrudecta, Foraminella ambigua* and *Lecanora chlarotera* appear	About 60
'Clean' zone (6–10)		Wide range of species found all over tree.	
	6	*Parmelia caperata* found at least at base with *Graphis elegans* and other species of *Parmelia* and *Pertusaria*	About 50
	7	*P. caperata* extends up the tree and species such as *Usnea subfloridana* and *Pertusaria hemisphaerica* appear	About 40
	8	*U. ceratina, Parmelia perlata* or *P. reticulata* appear	About 35
	9	*Lobaria pulmonaria, L. amplissima, Pachyphiale cornea, Dimerella lutea* or *U. florida* present, or crustose lichen flora well developed, there may be over 25 species per tree	Under 30
	0 1	*L. amplissima, L. scrobiculata, Sticta limbata, Pannaria* species, *U. articulata, U. filipendula* or *Teloschistes flavicans* present and crustose/ fruticose lichen flora well developed on twigs	Under 10

Note: For trees with moderately acidic bark in England and Wales

Source: Hawksworth and Rose (1976); Richardson (1992)

whether it is discharged at one point or over a wider area, and the characteristics of receiving waters (volume, flow rates, ecology, etc.) are all important in determining the impact on wildlife. Thermal pollution (the addition of warm water previously used for cooling by industry) lowers the oxygen content of water, especially in still waters. Low oxygen concentrations reduce the number of animals which the habitat can support. Industrial wastes may

Table 3.3:   Groups of aquatic pollutants in urban systems

Characteristics	Examples
Natural compounds with slight toxicity and slight ecological effects, but no bioaccumulation	Chlorides, sulphates and suspended soils
Natural compounds with slight toxicity and noticeable ecological effects, but no bioaccumulation	Phosphates, nitrates, sewage solids and thermal pollution
Natural compounds with toxicity and/or bioaccumulation plus slight persistence and/or mobility	Cyanides, nitrites and ammonia
Natural compounds with toxicity and/or bioaccumulation plus substantial persistence and mobility	Heavy metals, mineral oils and aromatic hydrocarbons
Artificial compounds with slight persistence and/or mobility	Some biocides and plastics
Artificial compounds with strong persistence and mobility	DDT and PCBs

Source: Hengeveld and de Vocht (1980/81)

Table 3.4:   Selected invertebrates characteristic of waters with differing amounts of organic pollution

Clean ⎯⎯⎯⎯⎯⎯⎯⎯⎯⎯⎯⎯⎯⎯⎯⎯⎯⎯⎯⎯⎯⎯⎯⎯⎯⎯▷ High organic pollution

Stoneflies (most species)	Stoneflies (a few species)	Molluscs	Fly larvae
Mayflies (most species)	Mayflies (some species)	Water hoglice	Worms
	Amphipods	Leeches	
	Caddis flies	Alderflies	
	Molluscs		

Source: Hellawell (1986)

Table 3.5:   Tolerance of some British fish to pollution

	Very sensitive	Sensitive	Tolerant
Indigenous species	Bullhead	Bleak	Common Bream
	Burbot	Pike	European Eel
	Chub	Roach	Perch
		Rudd	Silver Bream
		Ruffe	Tench
		Smelt	
Introduced species	Orfe	Pikeperch	Common Carp
			Crucian Carp

Sources: Hengeveld and de Vocht (1980/81); Maitland and Lyle (1991)

contain heavy metals and other toxins which kill intolerant species. Organic pollution from sewage and other organic waste increases microbial activity, placing a high demand on the oxygen present and reducing the diversity of organisms present. Suspended material (e.g. from sewage or mineral workings) reduces the light available to plants (thus reducing oxygenation) and blocks feeding in filter-feeding animals. Urea salt de-icers used by airports elevates ammonia levels in adjacent streams, killing some fish and benthic invertebrates such as *Gamerus pulex* (Turnbull and Bevan, 1995). Although individual pollutants are potentially harmful enough, mixtures may be disastrous. Sublethal levels of one pollutant may be fatal for animals suffering from stress caused by another pollutant, for example thermal pollution reduces the tolerance of some organisms to the toxic effects of heavy metals. Birds may also be affected by freshwater pollution, with some species having fewer territories where water quality is low, possibly due to effects on invertebrate prey (Rushton *et al.*, 1994). In Britain, pollution of one form or another threatens the survival of the ten most vulnerable species of freshwater fish, and has led to the extinction of two species (Maitland and Lyle, 1991).

The effects of pollution include changes in the communities of vascular plants, macroinvertebrates and fish (often causing a reduction in diversity). The reduced invertebrate diversity with organic pollution has been used to develop indices for monitoring such pollution in freshwaters. One such, the Trent Biotic Index (Hellawell, 1986), allocates a score depending on how many groups of invertebrates are present (more groups raises the score) and whether species from particularly intolerant groups (such as stoneflies and mayflies) are present (presence of intolerant groups increases the score). The overall index pro-

vides an objective view of the level of organic pollution present.

NOXIOUS SPECIES

When plants or animals are introduced into a new area, they may be very successful due to an absence of predators and competitors. If they are considered hazardous to humans or a problem to native wildlife, they are often thought of as noxious. However, if we examine the distribution and ecology of these species more closely, the pest potential of some such species may be less clear cut than was originally thought.

Noxious Plants

Several plants persist in urban areas either by virtue of their ability to exploit human-influenced habitats, or because they have been transported by humans. Several of these are of interest in studying the ecology of colonisation (e.g. Oxford ragwort) and cause little, if any, harm to native floras or faunas. Others are rampant and aggressive competitors which may overshadow native plant communities, reducing the diversity of the habitat. Many such species are abundant in disturbed sites. For two of these, Japanese knotweed (Plate 3.2) and giant hogweed, it is an offence under the Wildlife and Countryside Act (Department of the Environment, 1981) to promote their growth in the wild.

Japanese Knotweed

Japanese knotweed (Species Box 3.3) is a tall, vigorous perennial, which has become widespread in Britain, especially in lowland areas (under 200 m) and in the west of England and

Plate 3.2: Japanese knotweed growing at the edge of a bonfire plot

in Wales. It is likely to continue to spread as disturbed land increases, although its northern spread is probably limited by the length of the growing season and minimum temperature (Beerling *et al.*, 1994). Its dense canopy and persistence of stem litter may prevent other plants from growing underneath, reducing species diversity which in some cases weakens river banks. It is identified as a threat to conservation, especially in nature reserves in Wales and parts of the south-east and north-west of England (Beerling *et al.*, 1994). Although it should not be encouraged, and in some areas needs to be controlled, Japanese knotweed may be somewhat misunderstood (Gilbert, 1994). There are some plants which, in certain conditions, colonise stands of Japanese knotweed (Gilbert, 1992c). For example, below the flood level on the River Don in Sheffield, where stem litter is periodically removed during high water flow, there are stands containing lesser celandine, wood anemone, ransoms and bluebells. All of these species are early flowering and utilise spring sunlight before the knotweed canopy becomes too dense in mid-May. These conditions are not unlike those in deciduous woodlands. Stands above the flood level are only really colonised by hedge bindweed. In addition, a few insects feed on the leaves, including several bugs, beetles (both weevils and leaf beetles), and caterpillars of moths (the brick and Hebrew character) and butterflies (Beerling *et al.*, 1994). If colonisation by spring flowers and utilisation by some insects is a general rule, the necessity to control Japanese knotweed may need to be reassessed for areas such as urban derelict sites which are often fairly low in natural floristic diversity.

Giant Hogweed

Contact with giant hogweed sap causes reddening and irritation of the skin, followed by the formation of large irritating blisters, which subside after a few days, but may leave brown patches lasting several months. These blisters are caused by toxic substances (furocoumarins) within the sap which are activated by

Species Box 3.3: Japanese knotweed

Japanese knotweed is a perennial plant growing up to 3 m tall with hollow stems tinged with red, broad triangular leaves and greenish-white flowers. A native of Japan, North China and Taiwan, it was introduced into Britain in 1825 as a garden plant and fodder for livestock, and was a naturalised alien by 1886. It was found wild in London by 1900, near Exeter by 1908, Suffolk in 1924, West Yorkshire in the 1940s, Northumberland in the 1950s and by the 1960s was present across the country from Land's End to the Isle of Lewis. Only a small amount of seed is set in Britain, most being hybrid and inviable, and no seedlings have been found. Large (often several metres in diameter), monoculture clumps are produced by rhizomatous growth. Dispersal is by movement of rhizomes (or parts of rhizomes) in water along river banks and in soil (with inadvertent human assistance) in urban waste areas. Japanese knotweed is found mainly in relatively productive (and often artificial) habitats including road verges, alongside railway lines, wasteland, spoil heaps and river banks as well as in some woodlands.

Sources: Stace (1997); Beerling *et al.* (1994); Grime *et al.* (1988)

exposure to sunlight. The toxins are at their highest concentrations in the fruit and leaves, and are particularly high in leaves during spring and early summer (Lagey *et al.*, 1995). Children who play within clumps, or with stems cut from the plant, are most at risk. The plant appears less toxic to animals and is eaten by cattle, pigs, sheep and goats, although there are a few reports of ulcer formation on some animals.

Giant hogweed was introduced into Britain as a garden plant in the late nineteenth century from the Caucasus Mountains in Asia. Since then it has spread rapidly, especially in the last twenty years, and is now a pest, growing mainly along river banks but it is increasingly common in urban areas and along railway lines and roads (Tiley *et al.*, 1996). It is particularly common in lowland areas, especially in the east. This biennial or perennial species grows from a deep tap root to heights of 2–5 m and forms white umbels in June and July which produce 5,000 to over 50,000 seeds per plant (Tiley *et al.*, 1996). Wind disperses the seeds over short distances (up to 10 m), although the slipstreams of cars and trains enhance wind dispersal along roads and railways. It is in moving water that long distance dispersal takes place. Giant hogweed usually colonises a site initially as a single plant. Large groups (commonly up to fifty plants per group) quickly form as a result of short distance seed dispersal. Populations occur for several kilometres along some river banks. In north-west England two major populations run for 33 km through Bolton to Salford along the Rivers Tonge, Croal and Irwell, and for 15 km along the River Bollin from Hale to the Manchester Ship Canal. The shade produced by giant hogweed inhibits other plants and thus reduces river bank stability. In woodlands, giant hogweed accentuates the effects of tree cover, reducing ground vegetation still further. However, in grasslands, it may actually increase plant diversity by providing shade for other species. Single giant hogweed plants may stand in diverse vegetation communities, and even dense clumps can have several plants (mainly good competitors such as common nettles, cleavers and spear thistles) growing within them. In addition to generalist feeders such as slugs and snails, over thirty herbivorous insects (including true bugs, two-winged flies, moths and beetles) feed on the sap or leaves, with a similar number on the flowers (Tiley *et al.*, 1996).

Giant hogweed is therefore not universally damaging to environments, although its recent rapid spread, its effects on ground floral diversity, river bank stability and its toxic sap, mean that control is necessary at least in some areas. There are several control programmes in place by organisations such as the Environment Agency especially in the north west of England. Despite this, and restrictions placed on its introduction into the wild (Wildlife and Countryside Act: Department of the Environment, 1981), it was available for sale from garden nurseries as recently as 1992 (Tiley *et al.*, 1996).

Indian Balsam

Indian balsam was first introduced into England as a garden plant in 1839 from its native region of the Himalayas (giving rise to its alternative common name, Himalayan balsam). It is the tallest annual herbaceous plant in Britain, growing up to 2 m. Many seeds (800 per plant) are released explosively when the fruit opens, usually dispersing about 2 m. They are often carried much further by water and when soil is transported. It has been a pest in some habitats for nearly 100 years, and is still spreading, although its potential spread may be limited by the length of the growing season (Beerling and Perrins, 1993). It is a vigorous plant, competing with perennials such as common nettles (Prach, 1994). In some areas (e.g. parts of south Wales) it is common along river banks in stands of over 100 m in length. Few insects are confirmed as feeding on Indian balsam (Beerling and Perrins, 1993). Those that do include two species of aphid

(*Aphis fabae* and *Impatientinum balsamines*) and the larvae of the elephant hawk-moth. The aphids provide food for hoverfly, lacewing and ladybird larvae, adult ladybirds and predatory bugs (e.g. *Anthocoris nemorum*). Other species (including five species of bumblebee) use it as a major nectar and pollen source (Gilbert, 1992c).

The National Rivers Authority (now the Environment Agency) issue guidance notes for the control of Indian balsam specifically aimed at plants near watercourses (National Rivers Authority, 1994). Although a pest in woodlands and fens (where it forms dense stands which swamp other vegetation), some researchers consider that it is not a problem in bankside habitats. Indeed, it has been stated that Indian balsam brightens up otherwise dreary canal banks (Williamson, 1996). Usher (1986) recorded it as neutral for conservation management, suggesting that it may be more likely to establish where the native species were in decline due to pollution, than in undisturbed habitats. In Central Europe, Indian balsam is less of a threat than are Japanese knotweed or giant hogweed (Pyšek and Prach, 1995), especially along river banks with low conservation value. Since the species is still spreading in Britain and is an acknowledged pest in woodlands, it may still need to be controlled. Certainly it should not be planted, although it is not (unlike Japanese knotweed and giant hogweed) covered by the Wildlife and Countryside Act (Department of the Environment, 1981).

Rhododendron

Rhododendron punticum grows to 3–4 m in height, over a wide range of temperatures. It regrows quickly after cutting, is a prolific seed producer and disperser and is shade tolerant. It was introduced into Great Britain from Turkey in the late 1700s as a garden shrub and was widely planted in woods as cover for game (Becker, 1988). Since it does not establish where there is dense ground cover of plants or leaf litter, disturbance assists its regeneration. It is unpalatable to herbivores (producing andromedo-toxin), and few pests, grazers or diseases restrict its growth. The combination of widespread planting, regeneration in disturbed sites and unpalatability to animals are important factors contributing to its success. Given time, rhododendron replaces the understorey of woods with single species stands, casting heavy shade (large bushes only allow as little as 2 per cent of light through) and eliminating the flora in the ground (including fungi, lichens and mosses), herbaceous and low shrub layers, especially in acidic and disturbed sites (Usher, 1986). Since it has few associated insects or epiphytes, a rhododendron-dominated habitat has a low species diversity. Although birds and mammals use rhododendron as resting and breeding sites, native shrubs additionally provide food, and there are fewer birds in woods with rhododendron. The nectar is toxic and the leaves poisonous, resulting in a small insect fauna. In general, this species should not be introduced and control is often necessary, especially where it has become dominant.

Sycamore

Sycamore (Species Box 3.4) grows well throughout Britain and is difficult to destroy. Its seeds are wind-dispersed and it can establish at the expense of native trees, forming secondary woodland especially on industrial sites. It is invasive, casts heavy shade and dominates some mature woodlands, below which there is little regeneration (Peterken, 1993). Sycamore is reputed to be a poor habitat for wildlife (e.g. Rackham, 1986) and due to its

···

Species Box 3.4: Sycamore

Sycamore is a deciduous tree which flowers from April to June and sets seed from September to November. The leaves are five-lobed and toothed around the edges. It produces numerous wind-dispersed winged fruits which germinate in spring. Sycamores grow up to 30 m and are relatively frost-tolerant. Originally a native of central and southern Europe, sycamore was first mentioned as a garden tree in 1578, although it may have grown in Britain earlier, possibly having been introduced by the Romans. It expanded its range in the eighteenth century when it became fashionable to plant it, and is now one of the most widespread trees in Britain, being found in a wide range of habitats. The saplings can live in both shade and in the open. In urban areas seedlings are regularly found on abandoned industrial sites and even as a weed in gardens.

Sources: Harris and Harris (1991); Grime *et al.* (1988); Edlin and Mitchell (1985)

···

highly invasive nature, is considered a major conservation problem, although it is uncertain that long lasting single-species stands are produced unless they are planted as such (Fuller and Peterken, 1995).

Benefits to wildlife have also been described (Harris and Harris, 1991). It speedily adds nutrients to the soil because the leaves contain bases and rot down quickly. The prolific seed production provides food for birds and other animals during the winter. It supports a rich epiphyte community, with many lichens (194 species), liverworts and mosses growing on its bark. Bees use its nectar in May and forty-three species of insects are associated with the tree (Kennedy and Southwood, 1984). Introduced plants generally support fewer species of animals than do native plants, and sycamore has only about 10 per cent of the number of insect species found on some native trees such as willows (450 species), oaks (423) and birches (334). However, it has similar numbers of insect species to the native hornbeam and field maple (both with fifty-one species) and many more than the introduced sweet chestnut (eleven species) and horse chestnut (nine species). In addition, some insects are present in large numbers (e.g. the sycamore aphid, *Drepanosiphum platanoidis*). In fact, the biomass of insects on sycamore (35.8 g m^{-2}) can be larger than that on native trees such as oak (27.8 g m^{-2}) or ash (11.2 g m^{-2}), thus providing a useful source of food for insectivorous insects and birds (Mabey, 1996). Sycamore may therefore not be as problematic as is often suggested. Although it needs to be controlled in ancient woodlands where it restricts the growth of semi-natural trees (Fuller and Peterken, 1995) and should be removed from reserves, it may be a useful addition to some new woodlands (Peterken, 1993) especially in urban areas.

Noxious Animals

Some animals are considered problematic in carrying disease (e.g. rats and mice), or being an environmental nuisance (e.g. feral pigeons and cockroaches). Nowhere are these problems more apparent than in urban areas.

Rats and mice associated with buildings and other urban habitats are usually viewed as pests. Rats (Species Box 3.5) in particular are carriers of disease (including Leptospirosis which is occasionally fatal in humans) and thought to be indicative of dirty conditions. They also damage stored products. The house mouse is the commonest mouse in buildings, although it also lives independently of humans. Despite popular belief, house mice in Britain carry few diseases. They are more harmful to stored products, through spoilage while eating and leaving droppings. Although all three rodents are pests, their impacts on people are much less severe in Britain than in tropical countries, where they are major carriers of diseases such as typhoid and plague. Control is easier to achieve for animals living within buildings.

Feral pigeons (Species Box 2.20 and Plate 3.3) are classed as mainly environmental pests; blocking gutters with nesting material, marking buildings with droppings, and feeding on foodstuffs in grain stores. In common with many other birds, feral pigeons carry several diseases, some being hazardous to humans (e.g. psittacosis and encephalitis), and are most easily transmitted through contact (for example by hand feeding). Pigeon ticks (*Argas reflexus*) occasionally invade houses and bite humans, especially in spring and autumn (Dautel *et al.*, 1994). The ticks can cause skin

reactions similar to mosquitos and allergic reactions. Control of pigeons is difficult because rapid replacement occurs when animals are removed (Sol and Senar, 1995). The problems caused by pigeons are undoubtably exacerbated by deliberate feeding by people; one of the abiding city scenes is of huge flocks of pigeons surrounding tourists at the foot of Nelson's Column in London.

The three most common species of cockroach (German, Oriental and American, see Species Box 2.18) all inhabit warm places in hospitals, bakeries, restaurants, factories and private homes. Cockroaches are often perceived as harmful because of the tendency to form large populations, their odour and the fear that they will contaminate food and spread disease. They may damage food and other organic products such as paper and textiles, mainly by the deposition of faeces. However, they tend not to cause substantial destruction and there is little evidence that they transmit disease.

Managing Noxious Species

As we have seen, decisions over whether it is necessary to control noxious species are not always as straightforward as might initially be supposed. Assuming that control is necessary then the practical problems can be enormous. Four of the plants discussed above (Japanese knotweed, Indian balsam, rhododendron and sycamore) are amongst the 'top twenty' alien plant species in Britain, based on their distribution and abundance (Crawley, 1987). Several characteristics enhance this status; fast spreading (often by human agencies), few natural enemies and high reproductive rates. As such, prevention by legislation of the distribution or cultivation of noxious species is perhaps the most obvious starting point; two of

....

Species Box 3.5: Ship and common rats

Two species of rat occur in Britain, the common rat and the slightly smaller ship rat. The common rat grows up to 280 mm (head and body) with a proportionately slightly shorter tail (less than the head and body combined) which is often dark above and pale below. It has relatively smaller eyes and ears which are furry. The ship rat is up to 240 mm long (head and body) with an entirely dark tail which is usually longer than the combined head and body, and almost hairless ears. Ship rats live in male-dominated groups, while common rats live in colonies comprising several clans of breeding pairs or males plus their harems. Both species are predominantly nocturnal, although activity during daylight may occur with lack of disturbance, scarcity of food, or diurnal food opportunities. Although both are omnivorous, the ship rat is more vegetarian than the highly opportunistic common rat. Breeding in the ship rat occurs from mid-March to mid-November with three to five litters per year, each of around seven young. Common rats breed all year round producing seven or eight young per litter which are weaned at three weeks and females can reproduce at eleven weeks. Where conditions are less favourable, common rats restrict their breeding to the summer and autumn. Between them, these two species have interacted with human communities for centuries. The ship rat was introduced early, probably during Roman times, and has been largely replaced by the common rat which arrived in the early to mid-1700s from Russia. The disappearance of the ship rat from many cities (e.g. Manchester, Liverpool, Glasgow) has coincided with the demolition of dockside seed and flour mills. Its habit of living in buildings makes it easier to eradicate than the common rat which is more likely to be associated with urban waterways, refuse tips and sewers. Currently the ship rat is endangered with its population down to around 1,300 animals (750 in England, 550 in Scotland and none in Wales), whereas the common rat population is approaching 7 million animals across the country.

Sources: Harris *et al.* (1995); Corbet and Harris (1991)

....

the species described above (Japanese knotweed and giant hogweed) are covered by the Wildlife and Countryside Act (Department of the Environment, 1981) in this way. Control once a species is established is much more difficult. Their persistence and potential reproductive expansion provides the pest with the advantage in most cases. Methods include uprooting

Plate 3.3: Flock of feral pigeons in an urban park

stems by pulling or winching them out at the base, cutting, grazing, and the use of non-persistent herbicides such as glyphosate (see Table 3.6). In the case of animals, trapping may be employed (feral pigeons, rats and mice) or poisoning with baits (feral pigeons, rats, mice and cockroaches) or toxic fumigants (cockroaches). In many cases control simply suppresses populations rather than eradicating them, and decisions must be made about whether the problems justify the time and finance involved.

Some species cause concerns about environmental health issues; the cases of giant hogweed and Leptospirosis described above are obvious examples. Other species are implicated in the spread of diseases. Gull populations in towns are often infected by pathogens including salmonella (Furness and Monaghan, 1987). Around 10 per cent of the population is usually infected, although this can be higher in winter when landfill sites are more frequent feeding grounds. Higher salmonella incidence in gulls is often mirrored by increases in human infection, although it is difficult to identify cause and effect, since gulls may obtain the pathogens when feeding on human waste such as garbage and sewage (Fricker, 1984). Only small amounts of pathogen are carried by gulls and for only a brief period; they do not become actively infected. However, livestock drinking from small water bodies used by infected gulls may be susceptible and large gull roosts on reservoirs increase the costs of water treatment. Similarly, in the USA urban waterfowl populations carry botulism (Figley and VanDruff, 1982). This is caused by a lethal neurotoxin produced by a bacterium which lives in warm mud (especially where shallow areas dry out during hot weather). Mammals also carry diseases which can infect humans. In recent years public concern has focused on the transmission of *Toxocara* from domestic dogs to children, a parasite which can be damaging to health and sometimes causes vision problems. The eggs of *Toxocara canis*, carried by dogs, are more frequent in parks in Holland than those of *T. cati* which is carried by cats (Jansen *et al.*, 1993). There was no significant reduction in the incidence of

Table 3.6: Potential control methods for some problem plants

Species	Control method			
	Pulling/digging	Cutting	Grazing	Herbicide (usually glyphosate)
Giant hogweed	Pulling up young plants in April or May controls isolated, small individuals. In larger plants, the tap root is too large and can be dug out whole or severely damaged below the base of the stem.	Smaller plants (under 1.5 m) are cut below ground level, every two weeks in spring. This is repeated annually until no seedlings appear.	Grazed using cattle, sheep, goats or pigs from mid-March to the end of the growing season, but it needs many years to eradicate the plant rather than just suppress its growth.	Applied in March/April with a second application, if required, at the end of May. Could take from 5–10 years.
Indian balsam	Removal before the flowers form is important (to avoid seed production and dispersal) and may control small populations.	Cut at ground level before flowers form at end of June, but not too early or regrowth forms flowers with an increased number of seeds. This is repeated annually until no seedlings appear.	Grazed using cattle or sheep from mid-April to the end of the growing season, but intensive grazing may be needed until no new seedlings appear.	Leaves and stems are treated in late spring (before the end of June). This must be before flowers have formed and could take 5–10 years.
Japanese knotweed	Mature stems are uprooted from the base during June and July, and as much of the root as possible dug out. Will take at least 5 years.	Cut using a brush cutter, flail or grass mower every two weeks until the end of the growing season. Can take up to 10 years until no new growth appears.	After removing previous year's dead stems, young shoots are grazed using horses, sheep and goats from February to July continuously. This will only suppress growth, not eradicate the plant.	Applied in May when the plant is 1–1.5 m tall. Regrowth can be resprayed in September before plant dies back in the autumn. This needs to be repeated for 3–4 years.

Table 3.6: Potential control methods for some problem plants (contd)

Species	Control method			
	Pulling/digging	*Cutting*	*Grazing*	*Herbicide (usually glyphosate)*
Rhododendron	Roots are winched out if soils are light and the density is low or moderate. Top growth (above 1 m) being cut and burned first. Annual weeding of seedlings may be needed for at least 8 years.	Top growth (above 1 m) is cut and burned. Stems are cut to 100–300 mm above ground and used as firewood. The stems can be drilled and the holes filled with saturated ammonium sulphamate, regrowth being treated with glyphosate the following May to July.	Unpalatable to herbivores.	Stump regrowth is treated with herbicide directly in July to September. Stumps can then be treated between November and March.
Sycamore	Saplings and seedlings are uprooted.	Mature trees are felled and the cut stumps painted with ammonium sulphamate.	Grazing may suppress the establishment of saplings.	Stump regrowth is treated with herbicide.

Source: Becker (1988); National Rivers Authority (1994); Pycraft (1994); Tiley *et al.*, 1996

Plate 5: Roadside vegetation

Plate 6: Railway trackside habitats

Plate 7: Wildflower meadow planted near a housing estate

Plate 8: Vegetation colonising a cleared demolition site

T. canis in parks from which dogs were formally excluded. In all parks, *T. cati* eggs were usually present in sand boxes, where *T. canis* eggs were rare. Since children playing in sand boxes are more likely to come into contact with eggs than when playing elsewhere in the park, *T. cati* is possibly more problematic than *T. canis*.

HABITAT MANAGEMENT

Urban greenspaces are managed for two major purposes: to maintain or enhance nature conservation value; and to enable use by human communities. Many habitats are naturally transient (grasslands are eventually colonised by scrub and ponds silt up and dry out) requiring maintenance simply to survive. Elsewhere, pollution or other problems require amelioration to encourage the establishment of wildlife. Increasingly, urban sites have multiple uses; not just as areas for nature, but also for recreation and education (Millward and Mostyn, 1989). This may cause conflicts, especially at sensitive times of the year when breeding or care of the young is taking place. Where urban greenspace exists it is often under pressure from heavy usage (parks and playing fields), extreme management (gardens and road verges), toxicity or other adverse conditions (e.g. industrial wasteland and spoil heaps). Despite these pressures, even these habitats could be improved for wildlife and increasingly many areas are managed more sensitively.

Ecological Guidelines

Habitat enhancement and creation has received much attention in recent years (e.g. Gilbert and Anderson, 1998; Baines, 1995; Baines and Smart, 1991 and Emery, 1986).

One recent initiative is the community forest programme which seeks to create new habitats (woodlands and associated habitats) which will integrate several uses, including nature conservation. Some other schemes are described in the case studies. Although detailed descriptions of habitat creation and enhancement techniques are beyond the scope of this book, ecological guidelines for the planning and implementation of such management (e.g. Barker and Graf, 1989) are outlined below with examples from urban areas.

Continuity

Existing habitats, especially those with a continuity of management, are often more diverse, relatively stable and support locally rare species. Old churchyards are examples of sites with great continuity of use, which often contain desirable habitats which need sensitive management to avoid conflicts with the primary purpose of the site. In general, recommendations for churchyard management (e.g. Burman and Stapleton, 1988) suggest less intensive approaches, especially to grassed areas. Wildlife can be promoted if grass is not fertilised or managed like a lawn, but less frequently visited areas treated like meadows and cut preferably twice per season or at most once a month. Avoiding cutting in May and June (even in those areas used most often) allows important meadow species to flower, while cutting grass short in October helps smaller species gain a head start the following spring.

Diversity

Habitat diversity is an important focus, especially when establishing new areas. Heterogeneity is increased by supporting a range of successional stages (especially pioneer communities) and the use of rotational management practices, both in time and space. Habitats such as ponds, being transient, show a range of successional stages, eventually silting up and, after a marshy phase, becoming more terrestrial. The natural cycle of drying out, developing extensive plant cover and building up silt and weed, can take a considerable length of time. All stages from open water to marshy areas, including dried out basins, have wildlife value. It is important that such transient sites are managed as unobtrusively as possible, where possible maintaining all stages in the process by the creation of new sites rather than by intrusive management of existing ones.

Rarity

Priority should be given to locally rare habitats and sites supporting locally rare species. Although species of international importance are usually given precedence over national, regional or local value, in urban areas locally important species tend to be more relevant. For example, water voles could be encouraged in more stretches of urban rivers by sympathetic management, especially by promoting emergent and bankside vegetation, together with constructing banks from soft substrates (Kirby, 1995). It is also important to protect habitats with intrinsically low diversities if they are fragile or rare, or if they support vulnerable species, for example breeding sites for frogs and great-crested newts.

Size and Connectivity

Wide road verges, the sides of both used and disused railway lines, and canals and unculveted rivers provide wildlife corridors between sites within towns and cities (see earlier section on geographical influences). On roads, the mixture of management, from highly intensive at the edge to relatively unmanaged further away, provides several microhabitats which increase the richness of flora and fauna. Following construction, verges are seeded, usually with perennial rye-grass mixed with other species including red fescue, smooth meadow grass, crested dog's-tail and white clover. Trees are often planted, although mainly on larger verges and away from the edge to prevent mature growth from obscuring the view of drivers. Wide verges provide a range of habitats (grassland, scrub and woodland) which are relatively undisturbed. With the loss of early successional grassland elsewhere in the country, wide verges and areas alongside motorways could be established and maintained to re-create such habitats (Morris et al., 1994). These could provide sites suitable for some of our more threatened butterflies, including the silver-spotted skipper and Adonis blue.

Reducing Pollution

Pollution can be reduced if the use of agrochemicals in management is lowered and pollution control mechanisms (such as oil and litter traps in ponds) are built into site designs. Golf courses are often heavily managed areas, but may provide diverse habitats if they are not intensively managed all over. Not only does the use of herbicides, fertilisers and other chemical applications reduce plant diversity on the course, there is an increased risk of surface run-off into adjacent water courses, exacerbated by the need to irrigate the resulting lush, fertile grasses (Linde et al., 1995). There is also an advantage from the golf perspective in reducing chemicals: in the long term, chemical applications cause a change in species composition from swards with fine-leaved bents and fescues to those, less desirable for golf, which are dominated by annual meadow-grass.

Intensity and Naturalness

It makes ecological sense to employ low-intensity management, which is timed to avoid harming species especially at vulnerable times (e.g. pupating, nesting, breeding, raising young). Disturbance can be restricted, and public usage managed away from vulnerable sites and species, using buffer zones. The creation of impermeable surfaces may be avoided by diverting surface waters and creating ponds and marshes where appropriate. Soft edges provide gentle ecotones, avoiding sudden habitat changes which can be barriers to movement. Even semi-natural habitats (meadows surrounding woodland) may restrict immigration and colonisation by some small mammals (Kozakiewicz and Jurasińska, 1989).

Using Suitable Species

Native, local species are usually more suitable for planting. A knowledge of natural colonisation is useful in managing newly created sites (Davis, 1986), since naturally colonising species provide an indication of species suitable for restoration programmes. Too frequently, woodland planting on waste tips fails, due to a lack of aftercare and choice of inappropriate species. Although vacant industrial and/or residential land colonises naturally (often with interesting results), even after five years little woody vegetation is present (Clemens et al., 1984). On such sites there may be few sources of seed and trees probably need to be planted early to avoid slow growth or die-back due to competition from perennial grasses and herbs. Species should never be introduced into an area, and only reintroduced to newly created

sites, or once the reason the species failed before has been remedied.

Management Planning

Because of the time-scales involved in site development, advance planning can substantially improve the end results. One good example of how early planning can result in a rich diversity of habitats is Mere Sands Wood in western Lancashire. Once used for gravel extraction, the site is now managed as a local nature reserve by the Lancashire Trust for Nature Conservation. Even while the site was being worked, the end use had been identified and management of the margins was proceeding. As far as possible, existing boundary woodlands were retained throughout the extraction process. Grading and sculpturing of edges, creating shallows and connecting waterways, islands and other physical features were put in place prior to abandonment. This gave a substantial start to the development of the site and, although subsequently many new features have been added, the initial planning process provided a sound baseline from which to progress.

Monitoring

Monitoring the success of management practices allows alteration of the management plan if necessary. Monitoring species can also identify deficiencies in habitats. For birds, common species of inner city areas are often ubiquitous, also living in urban open spaces and suburban areas (Hounsome, 1979). Zones can be drawn up based on the birds which would be expected in different areas (see Table 3.7). This 'league table' is useful in indicating habitat deficiencies in relation to missing species: a site with all species in groups 1 to 4 from Table 3.7 except mallards, may have an

Table 3.7: Zonation of urban birds

Highly ubiquitous ——————————————————→ Less ubiquitous

Group 1	Group 2	Group 3	Group 4	Group 5	Group 6	Group 7	Group 8
Feral Pigeon	Blackbird	Blue Tit	Carrion Crow	Coal Tit	Blackcap	Barn Owl	Collared Dove
House Sparrow	Dunnock	Greenfinch	Chaffinch	Coot	Bullfinch	Chiffchaff	Kingfisher
Starling	Magpie	Mallard	Great Tit	Jackdaw	Great-spotted Woodpecker	Cuckoo	Marsh Tit
	Wood Pigeon	Robin	House Martin*	Moorhen	Long-tailed Tit	Garden Warbler	Redpoll
		Song Thrush	Kestrel*	Spotted Flycatcher	Mute Swan	Goldcrest	Rook
			Jay	Tawny Owl	Pied Wagtail	Goldfinch	Sand Martin
			Mistle Thrush	Willow Warbler	Stock Dove	Great-crested Grebe	Sedge Warbler
			Swift*		Treecreeper	Green Woodpecker	Sparrowhawk
			Wren		Tufted Duck	Lapwing	Turtle Dove
						Lesser-spotted Woodpecker	Whitethroat
						Linnet	Willow Tit
						Little Grebe	Woodcock
						Little Owl	Yellow Hammer
						Meadow Pipit	Yellow Wagtail
						Nuthatch	
						Partridge	
						Pheasant	
						Reed Bunting	
						Skylark	
						Swallow	
						Tree Sparrow	

E.g. City outskirts ——————————————————————————→

Suburban park ——————————————————→

City park ——————————→

City centre ——————→

Notes: Sites containing species in any one group should also feature most of the species found in lower numbered groups

* Species with less predictable distributions within towns

Source: Hounsome (1979)

absence of open water; a site without blue tits could be short of appropriate nesting sites; and lack of greenfinches may imply an absence of suitable food-bearing trees and shrubs. Such information could help to plan the addition of appropriate habitat features, siting nest boxes or planting appropriate food plants.

Taking Opportunities

There are many opportunities for wildlife enhancement of urban areas (e.g. Plate 3.4), not only in terms of using as many sites as possible, but also in management for the short as well as the long term. If sites which may be redeveloped in the next few years are managed for wildlife now, at the very least the environment was enhanced for some time (e.g. William Curtis Ecology Park, see case studies). If the planned development does not occur, the site will already show some benefits: if redevelopment happens, some features (trees, ponds, areas of meadow) may be incorporated into the final design.

Reclamation and Restoration of Derelict Land

Of all habitat management processes, reclamation and restoration are perhaps the most extreme, since both involve the conversion of sites from one use to another. Although these terms are often used interchangeably, they do have separate and precise meanings. Restoration is the conversion of habitats, which have previously been lost to waste or derelict conditions, back to their original form and use. Reclamation, on the other hand, is the winning back of land to a different use or form than the original one (Plate 3.5). The end use following restoration and reclamation of derelict and neglected industrial land was often to agriculture. This policy was increasingly questioned (e.g. Bradshaw, 1989) partly because of expense and the lack of need for extra agricultural land. Increasingly, end uses including nature conservation are examined as possibilities. A range of problems associated with derelict land require remedial techniques to

Plate 3.4: Remnant wetland now used as an urban nature reserve

Plate 3.5: Reclaimed gravel pit

enable plant communities to establish. These are covered in detail by several texts (e.g. Bradshaw and Chadwick, 1980; Gemmell, 1977; and Harris *et al.*, 1996) and a summary of those from a variety of industrial and mining wastes is given in Figure 3.2.

The physical nature of post-industrial sites often inhibits plant growth. If soils are too compact, drainage is prevented and the soil becomes waterlogged. If waterlogging persists, soils become anaerobic; this both reduces the oxygen available to roots and promotes growth of micro-organisms that remove nitrates (which may reduce nitrogen availability for plants). Conversely, soils composed of loose material suffer from wind and water erosion and have poor water-holding capacities. Steep slopes may be unstable and dangerous: in 1966 a colliery spoil heap in South Wales slipped down a hillside killing 142 people. Vegetation establishment, through natural colonisation or planting, provides long-term solutions to both compaction and loose substrates, since roots break up compact material and bind loose structures, incorpor-

ate organic matter and improve water-holding. In the short term, vegetation cover can be encouraged by preparing the substrate; by ripping compact and waterlogged surfaces, landscaping steep slopes and incorporating organic material into loose soils.

Extremely acidic or alkaline substrates influence plant survival, either directly by damaging roots or indirectly by inhibiting some metabolic processes in the soil and thus limiting nutrient uptake and root respiration. Even when not extreme, low and high pH may reduce plant growth by influencing chemical reactions occurring in soils. Acidity (e.g. in colliery spoil and mining wastes) may increase the soluble forms of some metal ions (e.g. zinc, copper, lead, iron and manganese) to toxic concentrations. Some metals (e.g. iron and aluminium) reduce the availability to plants of nutrients such as phosphate by forming insoluble compounds (ferric phosphate and aluminium phosphate). The alkaline nature of power station ash, chemical wastes and blast furnace slag causes phosphates to produce insoluble calcium phosphate; a form

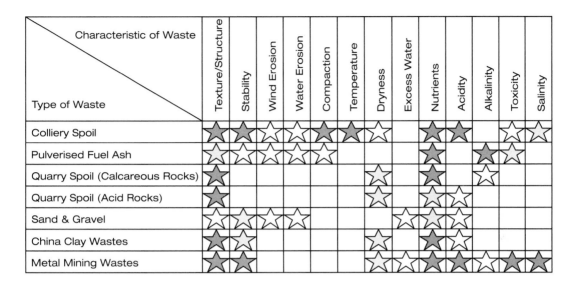

Characteristic may:

    be present and/or cause some problems

☆   cause moderate problems

    cause major problems

Figure 3.2: Major characteristics of industrial wastes which limit soil and vegetation development.
Sources: Bradshaw and Chadwick (1980); Gemmell (1977); Harris *et al.* (1996)

unavailable to plants. Some nutrients (e.g. boron, magnesium, iron and manganese) may be deficient at high pH levels, while others (e.g. aluminate and chromate) may occur at toxic levels. Natural leaching eventually neutralises some or all of the effects of extreme pH, but may pollute adjacent land and water. For some wastes (e.g. colliery spoil), natural leaching takes a considerable period of time. Additions of calcareous material (including crushed limestone or lime) may lower acidity somewhat faster. Alkaline substrates may be acidified using a variety of sulphur compounds depending on the waste involved (Gemmell, 1977).

Heavy metals (such as copper, lead and zinc) are toxic to plants even at low concentrations. Heavy metals occur in metal mine waste spoil heaps and smelter waste, as well as sewage sludge. Toxic effects include reductions in root, shoot and leaf growth. Natural leaching may reduce surface concentrations to a limited extent, but hundreds of years may be required before concentrations in lower soil layers fall to non-toxic levels. One possible solution is to cover waste with organic material, thus providing a growth medium for plants above the toxins. In addition, organic matter promotes formation of complexes of metal ions, thus reducing their availability and impact. Another solution is to use plants which tolerate toxic metals; species that do include creeping bent, common bent, red fescue, ribwort plantain and bladder campion. It appears that individuals growing on mine waste are tolerant, but those of the same species growing elsewhere are not (Bradshaw and Chadwick, 1980). This evolution of tolerant ecotypes usually results in tolerance only to the particular toxins found at the site; species on copper waste are not tolerant to zinc, and vice versa. Even when tolerant species are present, the fertility of metal waste tips is often quite low and plant growth is patchy, often being concentrated around pockets of nutrients. Other compounds also influence plant communities. Salt in some industrial wastes (e.g. colliery shale) reduces the capacity of plants to absorb water and so reduces growth. Natural leaching will eventually lower the concentrations, and this can be assisted by breaking up the surface to expose a larger surface area to winter rain.

Many derelict sites have poor soils, often as a result of the top soil having been removed. The remaining material may be very low in essential nutrients. Although this can inhibit the establishment of initial plant cover, low fertility assists long-term establishment of species-rich grasslands by reducing the capacity of more vigorous species to compete. In the short term it may be necessary to encourage growth by adding fertilisers, especially to address nitrogen deficiency. This is often only a temporary measure; the amounts of nitrogen required per year frequently being beyond the practical scope of fertiliser application. Sewage sludge and other organic materials help to prevent rapid leaching of nitrogen. Slow release fertilisers, although useful, are very expensive (especially for large areas). Long-term solutions involve the use of leguminous plants (such as common bird's-foot-trefoil, white or red clover and gorse) which have symbiotic relationships with nitrogen-fixing bacteria living in nodules on the roots. These bacteria, of the family Rhizobiaceae, convert atmospheric nitrogen to nitrates; a form accessible to plants. Similar associations occur between other bacteria and some non-leguminous plants, including alder, some

liverworts and a few shrubs and trees such as bog-myrtle. However, phosphorus may also be needed, especially since many nitrogen-fixing species are susceptible to phosphate deficiencies. Derelict sites may also be deficient in potassium, some macronutrients such as sulphur, calcium, magnesium and iron, as well as micronutrients (elements which are used in very small quantities by plants for a variety of metabolic processes: e.g. cobalt, zinc, copper and boron). Deficiencies of any of these will reduce growth.

Where plant communities are absent or sparse, low density planting of a nurse crop (usually an annual species which is not intended to be part of the final community structure) allows other species to establish by binding or breaking up soils, reducing heat, adding organic material and improving nutrient status. Invertebrates such as earthworms (Species Box 3.6) are important components for soil development. Depending on the soil type, inoculation of earthworms during reclamation increases the rate of growth and yield of plants, possibly by encouraging root growth (Edwards and Bohlen, 1996). However, their natural dispersal is slow and sites with high acidity support few earthworms (Davis, 1986).

Strategies for Nature Conservation

At a site level, management can increase wildlife value and reduce disturbance and vandalism. However, at a larger scale, strategies are needed to integrate nature conservation into other developments. The identification of ecological zones (areas of similar gross vegetation) may help to prioritise management, and many local authorities have carried out habitat surveys of their areas. Two types of survey are usually used: phase I habitat surveys are

Species Box 3.6: Earthworms

Most British earthworms belong to the family Lumbricidae, although some rare or introduced species from other families in the same suborder also occur. Only about twenty species are common in Britain, some of which occur near the surface of the soil, whilst others live much deeper. Adult sizes vary between species, but British earthworms range from less than 20 mm (*Dendrobaena pygmaea*) to over 300 mm (*Lumbricus terrestris*). Earthworms are roughly cylindrical, segmented animals with a mouth at the front end (covered at rest by a lobe from the first segment which also acts as a sense organ for tactile and chemical stimuli) and the anus at the back. They move by lengthening or shortening a series of muscles along the body segments while chaetae (bristles) on the middle segments grip the substrate. Earthworms feed mainly on decomposing plant material in the soil. Although they are hermaphrodite (each animal has both male and female reproductive organs) they usually cross-fertilise. They mate for up to four hours, lying head to tail, touching at the clitellum (the saddle); in some species being held together by the genital pads. In *Lumbricus terrestris* copulation takes place at night on the soil surface, whereas in most other British species it occurs below ground. Several egg cocoons containing up to twenty eggs are produced, but it is rare for more than one egg to develop to hatching.

Sources: Edwards and Bohlen (1996); Sims and Gerard (1985)

general descriptions of the habitats and vegetation types present, often limited to broad community types and species lists; phase II surveys are more quantitative and enable the communities to be compared to other systems such as the National Vegetation Classification. Important sites within a region may be identified by notification as SSSIs, or local designations such as SBI, SNCI or SINC (e.g. Sites of Biological Importance in Greater Manchester, and Sites of Importance for Nature Conserva-tion in Glasgow and Birmingham). Documents including area based strategies, biodiversity action plans and site management plans may also help to identify and hence prioritise conservation issues. There has been an increase in the production of nature conservation strategy documents over the last twelve or so years (H. Firman, unpublished). The earliest strategies were produced by Metropolitan County Councils and subsequently about two-thirds of Metropolitan District Councils have produced, or are in the process of producing, similar documents. Few at present are based on sound ecological surveys, although increasingly these are necessary for the production of Biodiversity Action Plans. The production and adoption of such strategies, particularly when based on ecological surveys, should provide a baseline for nature conservation practice. Biodiversity Action Plans, together with other environmental issues, are part of Local Agenda twenty-one programmes of sustainability resulting from the United Nations Conference on Environment and Development (held in Rio de Janeiro in 1992). Acknowledgement of the importance of urban wildlife and greenspace, together with appropriate inventories and strategies, should facilitate integration of nature conservation into the planning process.

Urban Wildlife and City People

The development of towns and cities results from the need for housing, work and other facilities for people. This puts considerable pressure on habitats and wildlife, although increasingly the benefits of wildlife for people are acknowledged. Certainly urban dwellers are keen to live near to greenspace and may even be active in its management (Emery, 1986). It is important that sites are close to residential areas; in many cases, people only travel short distances to greenspace, and even large areas rarely recruit visitors from more than 1 km away. This is particularly true of the distances that children travel (especially unaccompanied), and there is a recommendation that there should be at least 2 ha of greenspace within 0.5 km of any home (Harrison et al., 1995). There is no doubt that there are advantages in promoting urban wildlife for education, recreation and aesthetics.

Unfortunately there are also conflicts, and it is often necessary to carefully manage all the users of a site and educate the public about the impacts humans have on the environment.

Perceptions of Wildlife and use of Open Spaces

Local human communities like to make use of open spaces, often preferring a diversity of opportunities, rather than any one habitat type in particular (Johnston, 1990). In general, the public favour natural looking environments and will elect to live near to such if possible (Millward and Mostyn, 1989). In a comparison of Japanese and American perceptions of wildlife, both groups had strong affections for favoured species and enjoyed contact with aspects of the environment, although Japanese attitudes were more narrowly focused towards maintained rather than wild environments (Kellert, 1991). Public use of urban parks is dictated by local perceptions (Johnston, 1990), although the views of some communities from inner city areas (e.g. ethnic minorities and the young) tend not to be well represented in surveys regarding the usage of urban parks. The major motivations for site use are physical activity (exercise and relaxation) and enjoying nature (presence of a natural setting, looking at wildlife and enjoying scenery). The principal problems perceived are social conflict (personal safety and noise from other users) and maintenance (litter control and repairs). Tackling these problems may not necessarily be incompatible with the maintenance of a natural setting. For example, ground visibility can be increased, and naturalness retained, by reductions in shrub layers and raising tree canopies. Where visitor numbers are high, problems of disturbance and trampling occur, and zonation of use within and between urban sites may be necessary.

However, for many sites the reason for their creation and survival lies in the benefits gained by the public. Whilst management may enhance such sites, and zonation of use may prevent degeneration, it is important that local communities are involved and do not become distanced from their recreation sites.

Some wild animals (such as urban waterfowl) are enjoyed by a large proportion of the public from diverse backgrounds (Plate 3.6), and have aesthetic, educational and recreational potential (Figley and VanDruff, 1982). The use of some areas for recreation has certainly brought environmental issues to public attention. For example, the increasing use of open water for recreational activities has led to more people being aware of water quality problems, such as the toxic effects of some algal blooms (see earlier). However, even when sites are managed for wildlife, there can be problems in perception. For example, in order to encourage wildflowers on a previously frequently mown bank, management was limited to a single cut in autumn (Gilbert, 1981). After two years, when local residents were asked for a reaction, most stated that they preferred the previous management practice (mainly because it was tidier).

Environmental Education

If we are to ensure a future for urban wildlife, it is important that people are educated to understand something of the workings of nature and the implications of human impacts. Environmental education covers education in the environment, about the environment and for the environment (Palmer and Neal, 1994). Urban dwellers with relatively low exposure to nature, often have little familiarity with wild plants and animals. In a comparison of eleven-year-old children from rural Northern Ireland with similar children from Belfast, Williams and McCrorie (1990) found differences in attitude to the environment, with urban children having a generally lower environmental consciousness. They suggested that environmental education should be tailored to reflect the environmental background of the recipients.

Plate 3.6: Members of the public feeding wildlife in urban parks

a   b

Managed urban greenspaces often fulfill an educational role and urban ecology parks such as the one at Camley Street in London (see case studies) cater for thousands of school-children every year. In many cities, countryside managers in urban and urban fringe sites have found their roles expanded to embrace environmental education. In some cases this has also applied to city park staff. While appreciation of the environment can be taught in subjects as diverse as english, art and technology, ecology is the most obvious subject to use urban habitats. Natural processes can be investigated in urban greenspaces, human impacts on the environment examined at airports and factories, and treatment of pollution and wastes discussed at sewage works and water treatment plants (Fail, 1995). Environmental education within school grounds is also useful, and avoids the necessity to visit sites elsewhere. Pupils can have a role in habitat creation and management, to improve their own environment and gain a feeling of ownership. Where there is a lack of skill or ability for such management, help may be gained from parent groups and other schools, local colleges or universities. A successful programme of cooperation has been in place for five years between Pike Fold Primary School in North Manchester and students studying environmental subjects at the Manchester Metropolitan University. This has improved the wildlife value of the school grounds, provided environmental education materials (much of it using features of the school grounds) and hands-on experience of habitat management for both pupils and students. The National Curriculum in England and Wales now incorporates environmental education as a cross-curricular theme. Although there are problems with a cross-curricular approach, an awareness of the local environment and the increased use of it in education, may help to protect urban habitats and wildlife in the future.

4

CASE STUDIES

•

The following case studies provide a discussion of some opportunities, problems and solutions which exist within urban and urban fringe habitats. They are drawn from throughout Britain (Figure 4.1) and, although each is unique, they have implications for similar habitats elsewhere.

RECLAIMING LIMESTONE QUARRY FACES IN DERBYSHIRE

For over 2,000 years limestone has been quarried in the United Kingdom, first for building stone and later for agriculture, construction and industry. As the most economically important hard rock in Britain, demand is high and the amount extracted doubled between 1972 and 1987. Early extraction techniques produced small quarries which, when abandoned were often naturally colonised, and have subsequently become important for nature conservation (Plate 4.1). Limited nutrient and water availability, together with disturbance caused by rockfall and grazing, restricts the establishment of vigorous species and promotes early successional species-rich plant communities similar to semi-natural grasslands (Davis, 1982b). In many abandoned limestone quarries, the majority of plants are typical of the locality and are not specialists of such semi-natural grasslands (Hodgson, 1982). It is often only those disused for many years (perhaps over fifty years) where specialist plants are present in large numbers (Cullen *et al.*, 1998). This case study first examines some of the problems in the reclamation process and then focuses on one example of quarry reclamation.

Quarries were originally worked by hand and later using black powder blasting. These processes left low, ragged faces which eroded readily following abandonment. Modern quarries are much larger operations, up to 1 km across. The removal of huge swathes of countryside is not the only problem; the increased quarry depths, combined with modern explosives, produce tall, highly engineered quarry faces which do not erode easily (Plate 4.2). On abandonment, modern quarries take

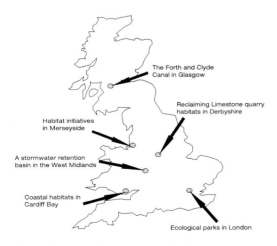

Figure 4.1: Location of case study sites.

Plate 4.1: Abandoned quarry

Plate 4.2: Modern production quarry

much longer to be colonised by wildlife, and the feasibility of successful reclamation is now an important consideration when planning permission is sought for new extraction sites (Department of the Environment, 1989). Several options exist for the reclamation of limestone quarries, depending on the planned after-use. Many are returned to agriculture and others used for forestry, recreation or nature conservation, although there is an increasing tendency to integrate several uses within larger sites. A common technique is infilling with landfill from domestic or industrial waste (see earlier).

Often it is not possible to fill the entire void left by quarrying and when quarry faces remain, reclamation is more difficult. Modern faces are sheer and artificial in appearance. Although natural erosion eventually fractures the stone to produce scree slopes and other more natural features, experience of older, smaller quarries where these processes take decades suggests that in modern quarries it may take hundreds of years. In addition, rockfall following erosion could prevent the site being used for recreation and amenity. Several techniques have been used which treat the face to reduce rockfall and disguise the engineered origins. In a few cases, pre-split smooth blasting (based on civil engineering techniques) has been employed to produce very stable faces with even more engineered surfaces, to reveal the underlying geology for scientific and educational purposes. Where more natural landscapes are desired, material is often pushed up to or over the quarry faces (a process known as backfilling). This material, often quarry waste, rock, or wastes from elsewhere (including domestic or industrial wastes), is usually covered by organic matter which may then be seeded. The organic cover increases moisture retention and provides a nutrient base in which plants can establish. For the first few years following treatment, many plants which do well are typical of fertile areas. With time, nutrients are leached from the organic material, rabbits graze the vegetation, and the species become more typical of daleside communities. This technique is also used following blasting to produce simple landscape features, such as scree slopes and shelves (Humphries, 1979).

Selective blasting, to produce landforms which are common on natural dalesides nearby, has been taken a step further at Tunstead Quarry in Derbyshire. A team of geomorphologists and ecologists have developed a process called landform replication which aims to produce ecosystems in limestone quarries similar to those on neighbouring natural dalesides. The process consists of two techniques: restoration blasting followed by habitat reconstruction. The former involves selectively blasting quarry faces to produce rock buttresses, headwalls and bare scree slopes; all typical geomorphological features of natural dalesides (Gagen et al., 1992). In the second stage, selected areas of scree slope are covered with fine limestone quarry waste, which is hydraulically seeded (i.e. seeds are suspended in a water-based slurry mixed with organic ameliorant) using a mix containing species found on nearby natural dalesides (Bailey et al., 1992). Hydraulic seeding prevents the need to use machinery directly on the scree slopes, reducing compaction and movement of the rock skeleton. Trees are planted in pits in similar positions to those on natural dalesides.

At Tunstead the establishment of plants on landform replications was monitored, and compared with neighbouring unmanaged disused quarries and a natural daleside. Initial results were encouraging, since almost all the species planted grew during the first season (Bailey, 1995). Subsequently however, some problems occurred. Although the species establishing on the reclaimed sites were more similar to the neighbouring natural daleside than they were to the disused quarries nearby, the coverage of vegetation was patchy (Wheater and Cullen, 1997). In addition, many species which grew in the year following seeding, were not found three to five years later (Cullen et al., 1998). Grazing by rabbits reduced plant cover, and exclosure experiments demonstrated that more diverse communities could develop in the absence of rabbits. Eventually, once the vegetation is established, grazing may prevent vigorous species from taking over and help to maintain species-rich plant communities. However, the actions of rabbits are also detrimental to the plant colonisation of other quarry habitats, such as quarry floors (e.g. Davis *et al.*, 1993). It may be necessary to make several introductions of plants over time as the communities develop (Humphries, 1980), and this together with exclusion of rabbits by fencing and early additions of nutrients might help to prevent bare ground problems at sites such as Tunstead.

Plants are introduced directly in such reclamation programmes. However, other organisms, such as invertebrates, must colonise naturally. Invertebrates are central to ecosystem development: woodlice and millipedes, for example, may be important in the development of soils in disused limestone quarries (Davis and Jones, 1978). Studying invertebrate diversity, including predators such as spiders and ground beetles, may give an indication of the relative stability of the ecosystem. The newly reclaimed sites at Tunstead Quarry were surveyed for invertebrates inhabiting the upper layers of the soil and soil surface (Wheater and Cullen, 1997). More plant and detritus feeders (millipedes, molluscs and woodlice) were found in older, more established quarries than in recently reclaimed areas. The predatory groups (spiders and ground beetles) gave similar results to studies on other types of newly colonised sites (e.g. Gilbert, 1989), with more surface-active predators being found on recently reclaimed sites. These predators may be feeding on insects such as flies which use bare areas as alighting, basking and grooming sites. Vegetation and invertebrate surveys at Tunstead show that, even when plant communities are encouraged and begin to establish, there is a delay in the recruitment and stabilisation of appropriate animal populations (see Figure 4.2). Seven

years after reclamation, these sites have not yet settled down. Frequently, reclaimed land is left to its own devices, or at best given five years of after-care. However, these studies suggest that these short time-scales are inadequate, and continued monitoring is necessary to refine the techniques employed.

ECOLOGICAL PARKS IN LONDON

The William Curtis Ecological Park was the first ecology park established in London (Emery, 1986), or indeed in Britain. In 1978 this small (0.8 ha) abandoned lorry park was developed into a site with several habitats (including areas of grassland, tree planting and wetland), providing not only nature conservation interest, but also environmental education value. Up to 15,000 visitors per year, many from local schools, used the site until it was redeveloped in 1985. Although temporary, the park demonstrated the range of wildlife which could live in well-managed open spaces surrounded by otherwise fairly inhospitable urban areas. Other more permanent ecology parks have since been created in London and other cities including Manchester, Birmingham and Sheffield. The development of one of these, at Camley Street in the London Borough of Camden, is described below.

Situated alongside Regent's Canal near to Kings Cross and St Pancras Stations, Camley Street Natural Park is a linear, 0.9 ha site comprising a mixture of habitats. The site, a disused coal yard which had been abandoned for twenty years, was purchased from British Rail in 1981 by the Greater London Council (GLC). Although originally planned as a coach park, a local wildlife group pointed out the diverse fauna and flora which had developed in the twenty years or so since

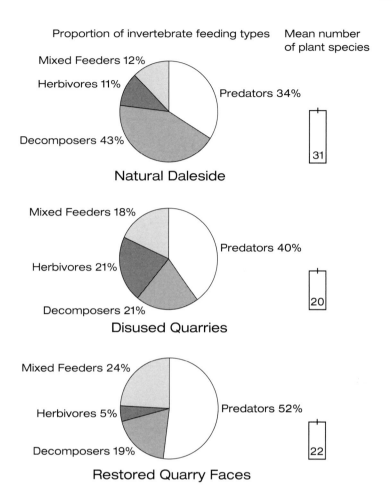

Figure 4.2: Numbers of plant and animal species in differing limestone quarry environments. Although planting vegetation can be successful (restored areas have reasonable plant establishment compared to the natural daleside and disused quarries – bar graphs), invertebrate communities may take longer to establish and initially be dominated by transitory predators (pie graphs).
Sources: Cullen (1995); Wheater & Cullen (1997); Cullen *et al.* (1998)

abandonment. The species present were typical of demolition sites (see earlier section) with some grassland areas, tall herbs, shrubs and trees comprising Yorkshire-fog, cocksfoot, common couch, Oxford ragwort, rosebay willowherb, butterfly-bush, goat willow and sycamore. It was proposed that the site would be valuable as an educational resource, and that the diversity could be increased by creating aquatic habitats (Plate 4.3). The pro-

posal to establish an ecology park was approved in 1982. Landscaping began in November 1983 and by the following spring schools were beginning to use the site, although the official opening was not until 1985. With the demise of the GLC, the land was eventually transferred to Camden Council who lease it to the London Wildlife Trust at a peppercorn rent. Further details of the development are given by Johnston (1990).

Plate 4.3: Wetland at Camley Street Natural Park

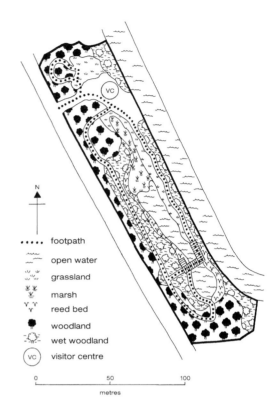

Figure 4.3: Map of Camley Street Natural Park.
Sources: Johnston (1990) and information from Camley Street Natural Park

Today, strips of woodland surround the site, screening it from the road and canal. The park includes open water, surrounded by reedbeds and marshy areas (see Figure 4.3), with areas of grassland which are managed as meadows (cut twice per year). The site is now rich in wildlife, with fifty-three species of trees and shrubs, over 260 herbaceous plants, and thirty-four grasses, sedges and rushes. Not all the plants were introduced, and some have colonised from elsewhere (e.g. water figwort, brooklime and lesser water-parsnip). Several waterbirds (e.g. coots and moorhens) nest at the edge of the pond, which also contains common frogs, common toads and smooth newts. The aquatic habitats are topped up from the neighbouring canal, which, although it has fairly high concentrations of some pollutants (e.g. heavy metals), is clean enough for relatively intolerant fish and aquatic larvae of mayflies and caddis flies. Eight species of dragonfly and damselfly have been recorded, including the ruddy darter and migrant hawker, which although not common nationally, are widespread in other ponds and lakes in London. Plants such as the butterfly-bush attract red admirals, peacocks and small tortoiseshell butterflies, whilst common blue butterflies occur on leguminous species such as birds-foot-trefoil. Common bird species on the site are those typical of many city parks (e.g.

blackbirds, song thrushes, blue tits, great tits and wrens).

Near the entrance is a visitor centre and classroom and several small areas planted to attract wildlife and provide a range of visual and olfactory stimuli. Three full-time members of staff, including one teacher and several volunteers, manage the park, which is now an important site for nature conservation in London (Waite *et al.*, 1993). The site is visited by around 13,500 visitors each year, mostly from London (90 per cent), especially from Camden (65 per cent), although 5 per cent are from elsewhere in the UK and the remainder from overseas. The range of habitats and number of species in this small site helps to make wildlife accessible, and it is therefore not surprising that much of its success is in environmental education. Over 3,500 school children visit on formal educational trips per year, while other educational initiatives account for nearly 1,000 more visitors. The school visits encompass aquatic ecology, food chains, ecology and conservation, different habitats, plants and invertebrates. Activities include pond dipping from boardwalks and collecting surface-active animals from under strategically placed logs. The other educational strand covers training courses, land use and development courses, landscape studies and ecology, photography and art and habitat creation and management.

Sites in the middle of an urban area are potentially very valuable for commercial development, and this park was recently threatened by an extension to Kings Cross Station (as part of the Channel Tunnel rail link interchange). The threat may have now been resolved and the siting of the nearby interchange may result in the loss of only a small section of the park. Despite the removal of this threat, the park has a disturbed short-term future. The building activity for the rail link

will create dust and noise which could adversely effect wildlife and the enjoyment of visitors. This is in addition to current problems of dust and litter from waste transfer stations situated opposite the park. Careful management of the park will be needed to ensure that this successful enterprise survives the existing and future impacts and continues to be an oasis for wildlife, a formal educational resource for nearly 5,000 people per year and an area of enjoyment for many more.

HABITAT INITIATIVES IN MERSEYSIDE

Within urban areas, there are many sites with little nature conservation interest or benefit to local human communities. Polluted water courses and areas used for fly-tipping are just two examples. In towns and cities it is important to manage sites around local people, not only to protect wildlife from disturbance, but also to encourage ownership, foster environmental appreciation and education, and provide amenities for local communities. In recent years several initiatives have improved the urban environment, some through the use of national bodies, and others utilising local expertise. In Merseyside, several organisations have been involved in urban site management initiatives, three of which are described here to illustrate the benefits of involving local communities in conservation.

Landlife

Amongst the organisations involved in urban site management in Merseyside, Landlife not only helps to reclaim derelict land, but also plants and harvests wildflowers in the most unlikely surroundings. Beginning life as the Rural Preservation Association in 1975, the organisation changed its name on its tenth anniversary to Landlife. Now over twenty years old, Landlife has been involved in many areas of creative conservation in the region. A wildflower trading arm was established in 1983 and a technical guide (now in its second edition) for wildflower landscapes has been produced (Lickorish *et al.*, 1997). Many sites in Merseyside have been seeded with wildflowers, both to enhance the aesthetics of the area and, in some cases, to provide a crop from which seed is harvested and then sold to help to support their other initiatives (Colour Plate 7). In some cases, mixtures of flowers are sown on building rubble, motorway waste or other harsh substrates. Successful planting of wildflowers on harsh substrates is not always straight forward. It is important that suitable species are chosen (often wildflowers grow best on infertile substrates, although cornfield annuals require fertile soils). In many cases some preparation is required to break up solid impervious surfaces, remove extensive weed cover and reduce fertility (for example by incorporating infertile substrates within the growing medium). Other initiatives include the use of recycled clothing to make wildflower blankets (peat-free seedbed material in which to sow wildflowers) and a recent project using old bank notes as compost.

By raising wildflowers on urban sites, Landlife does not attempt to reproduce rural habitats. However, their simplified approach to sowing and use of ecologically valuable species, together with appropriate substrates and management can lead to richer, sustainable habitats in the longer term. In addition, such sites provide pleasing environments for local people, as well as opportunities for a variety of animals which find the environment favourable. The resulting semi-farmed environments are less vulnerable to public access than are some nature reserves. Indeed, Landlife

encourage local involvement in the planning, sowing and planting stages, and through the organisation of special events around the wildflower areas. Creative conservation initiatives such as these should not replace habitat enhancement and the creation of areas with high nature conservation value, but can cater for local community needs and complement other urban site management initiatives.

Landlife's most notable achievement has been the development of the National Wildflower Centre. This has arisen from the experiences in developing wildflower areas in Merseyside, especially through projects such as the Knowsley Wildflower Project (a joint initiative started in 1990 between Landlife and Knowsley Metropolitan Borough Council) which resulted in over 18 ha of wildflower landscapes being created in the borough. The National Wildflower Centre will be a focus for creative conservation as well as a venue for community involvement and leisure activities. Training and education in urban regeneration techniques will also be supported as will research and information exchange. Public events (theatre, music and arts) will supplement the available parkland with wildflower landscapes, a visitor centre and other facilities for the public. It will also house a wildflower seed business and plant nursery and a living seedbank for British wildflowers. Funding from sponsorship, including the Millennium Commission, has ensured that this exciting project is fully functional by the year 2000.

Sutton Mill Dam

Operation Groundwork was established in 1981 to promote environmental action for the benefit of communities at a local level. By 1983, five Groundwork Trusts had been set up in the north-west of England, with the first in

St Helens and Knowsley (later extended to cover Sefton). There are currently over forty trusts, which attract funding from the Department of the Environment (for the first five years), local authorities, local businesses and the national Groundwork Foundation. In Merseyside, the local Trust has worked with local industries and communities to create and enhance urban sites, especially on derelict land. Examples include reclamation of colliery and power station waste tips, creation of a trout fishery in an abandoned clay pit and, the example given here, the restoration of a former mill dam.

Sutton Mill Dam was originally used to power a corn mill in the eighteenth century, and subsequently for some years to supply water for a chemical works until the dam silted up (Smyth, 1990). Following plans in the 1970s to use the site to dump waste, a local residents' action group was set up to object and put forward alternatives. The 6 ha site (Figure 4.4) was purchased in 1979 by St Helens Metropolitan Borough Council, who originally planned a reclamation scheme involving recreational uses. However, the local residents, who regularly use the site, preferred a more natural approach. As a result the council raised the dam wall as the first part of a plan to reflood the site and develop a wildlife park. The Groundwork Trust was approached to design and implement the next phase: to remove silt from the open water areas; supplement the vegetation with natural planting; and create a network of footpaths.

Details of the management of this site are given in a management plan produced by the Groundwork Trust (1990) and are summarised here. The succession to wet woodland was arrested by raising the water level using a causeway, to create a small holding pond above the main area of open water, and a weir at the far edge of the open water. This design

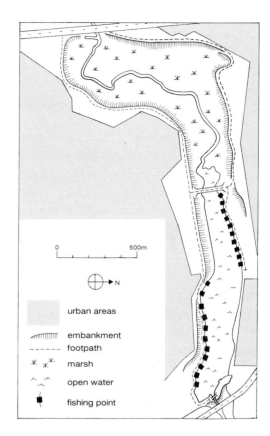

Figure 4.4: Map of Sutton Mill Dam.
Sources: Sefton, Knowsley and St Helen's Groundwork Trust

enables water to be retained in the main dam area (even in dry weather), yet prevents water backing up in very wet conditions. The upper reaches (about a quarter of the site) are now marsh (mainly reed canary-grass) with some willow scrub at the edges. The upper areas provide habitats for plants typical of emergent or wet conditions such as reedmace and yellow iris, together with their associated animals (including reed buntings). The water in Sutton Mill Dam is fed by Pendlebury Brook, which flows through both agricultural and urban areas before entering the reedbeds. Consequently the water is high in both organic matter and sediments, but is fairly well cleaned by the time it flows into the open water lower down. This is achieved partially by the cleaning properties of the reedbeds (see earlier) and partially by the holding pond where much of the remaining sediment settles out. It is anticipated that this pond will need to be dredged every five years or so.

The lake comprises about a quarter of the site and is up to 3 m deep. The dam was dredged to increase the depth of water and delay succession, but is still fairly heavily silted. The water quality is good and supports several species of fish including bream, carp and roach, with which the site has been stocked since restoration began. Each side of the open water is roughly divided into two, with half accessible to the public and the remainder protected for wildlife. A system of paths and boardwalks allows access to the waters edge and to the tops of the steep banks that surround the site. Fishing platforms are positioned along the accessible shores. The areas without access feature reedbeds and gently shelving banks, providing nesting sites for waterbirds, including mallards, coots and moorhens.

About a third of the site is wooded, including 0.25 ha of mature woodland. The existing mature woodland (mainly sycamore and willow) has been supplemented by an area about three times the size with native trees and shrubs planted from the mid-1980s. Along the path network are wildflower meadows and areas of rough grassland. All areas planted with native flowers, shrubs and trees have increased the floral diversity and provided habitats for birds, mammals and invertebrates.

Some vandalism (mainly of boundary walls, fences and gates) has occurred, and litter can be problematic, especially food wrappers from school children using the site at lunchtimes. There is some dumping in the shallowest and most accessible end of the open water and holding pond. The most obvious problem is fouling by dogs, since Sutton Bank Mill is a popular area for dog walking. However, the site has remained relatively free of the problems often associated with unwardened urban open areas. The early involvement and continued use by local communities may have contributed to the sustainability of the site. It is a continuing resource for nearby schools (educational material has been produced on several topics linked to the site) and for a local angling club who have the fishing rights on the dam.

Mersey Community Forest

At around 1,000 km², the Mersey Forest will be the largest of the Community Forests currently being developed in England. It was launched in 1991 with the long-term (thirty-year) aim of increasing the woodland cover across Merseyside and North Cheshire from 4 per cent to over 25 per cent (Mersey Forest, 1995). The planning stage took several years, with the Mersey Forest Plan approved in 1994 when the project moved into its implementation phase. In addition to increasing woodland cover, the project aims to provide a mixture of uses for the sites. There will be commercial woodland, areas which help to develop derelict land, sites for the conservation and protection of wildlife and areas where public usage will be encouraged. Plantings around residential areas will improve local surroundings and increase access to greensites. Leisure activities, both informal (walking, picnicking, cycling and riding) as well as more formal organised sporting events will be catered for. Art based projects (e.g. sculpture and plays) will be encouraged throughout the area. Environmental education will also be provided for;

woodland areas are potential open air class-rooms, many of which are readily accessible to local schools.

All of this is planned through a partnership of Mersey Forest staff together with nine local authorities (see Figure 2.1), and other organisations including the Countryside Commission and the Forestry Commission. Integral to the plan is the involvement of local industry and communities. To encourage involvement of industry, the project team offers advice on ways in which to increase planting on derelict, neglected and operational land, as well as on golf courses, school grounds, hospitals and landfill sites. The wide range of uses and proximity of sites to local people provides an impetus for local involvement. Communities, groups and individuals are encouraged to assist in planting, to participate (and organise) events within woodland areas, and to provide information about suitable sites which may be available for improvement. The enthusiasm with which the project has begun (so far around two million trees have been planted on over a hundred sites and over half a million pounds of European Union funding has been obtained to help encourage tree planting) suggests that the project is on its way to achieving its aims by the target year of 2025.

THE FORTH AND CLYDE CANAL IN GLASGOW

Increasingly, river or canal corridors in urban areas are managed for recreation and nature conservation (e.g. the Mersey Valley in Greater Manchester and the River Taff in Cardiff). These waterways, together with associated bankside habitats, often provide relatively long continuous areas of surprising wildlife diversity. The water itself, once pollution is controlled, may be host to a wide range of organisms, and important hydroseral communities are often present in adjacent areas, especially where periodic flooding has limited bankside development. These corridors are often important links between greensites in urban areas, and may even link rural and urban sites.

The Forth and Clyde Canal was built as a ship canal stretching about 60 km from the Firth of Clyde in the west to the Firth of Forth in the east, with a spur entering the centre of Glasgow. The canal passes through several urban areas, the most extensively urban stretch (about 15 km long) being through Glasgow and its environs (Figure 4.5). Started in 1768, the canal was completed in 1790, although it was navigable from the eastern seaboard over part of its length as early as 1773 (Bowman, 1991). Following financial losses from 1948, the canal was officially closed to navigation in 1963 and over the next ten years several sections were filled in, or otherwise blocked by developments including road building. Dereliction over the subsequent ten years led to the need to either fill in the remaining sections (several tragic drownings stimulated this call on safety grounds), or renovate the canal as part of a facelift of Glasgow (Davies, 1991). Some stretches were cleaned up over the following few years and several plans were considered to reopen yet more sections. In 1988, British Waterways proposed that much of the canal in Glasgow would be opened up.

The wildlife in and around the canal is rich, and sections in some of the most urban areas are locally important. The aquatic environment is reasonably large at 6–9 m wide (tending to be narrower in more urban stretches) and usually over 1 m deep at some point across its width. Floating plants (fat duckweed and common duckweed) and submergent species (Canadian waterweed, spiked water-milfoil, various-leaved pondweed, broad-

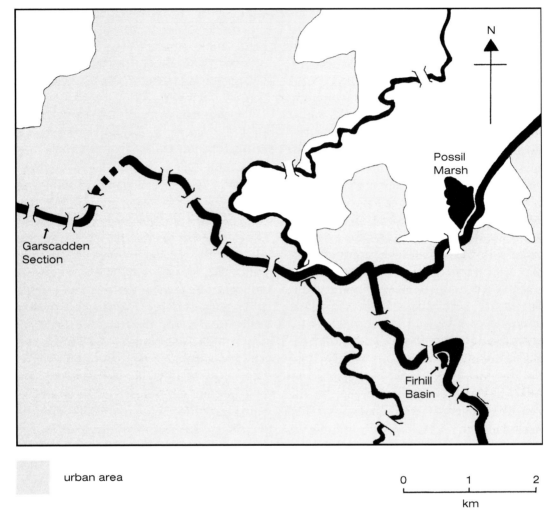

Garscadden
Section

Possil
Marsh

Firhill
Basin

N

urban area

0   1   2

km

Figure 4.5: Map of the Forth and Clyde Canal in Glasgow.
Source: Carter (1991)

leaved pondweed and small pondweed) grow in the open water. Within this region are also fish (pike, roach and perch) and various invertebrates such as leeches, snails, water beetles, water bugs and the larvae of damsel-flies, alder flies, caddis flies and two-winged flies. The large amount of bankside compared to open water provides habitats for a wealth of emergent plants, including water-plantain, reed sweet-grass, lesser water-parsnip, water-

cress, floating bur-reed, branched bur-reed, tufted loosestrife and meadowsweet. Within the shelter of these plant communities breed common frogs and several common water-birds (mute swans, mallards, coots, moor-hens). Other waterbirds, local to this part of the canal, include breeding populations of lit-tle grebe and occasional tufted ducks. A range of terrestrial birds are resident, especially the usual urban species of blue tits, wrens, blackbirds, chaffinches and magpies. Swifts often feed on insects over the water surface. Where the edges are severely landscaped, the bankside communities are the least interesting. Similarly, intensively managed towpaths provide relatively few habitats. Where the banks slope shallowly, sometimes due to siltation or bankside erosion, marshy conditions may prevail. Reed sweet-grass is dense in places forming a fringing reedbed, within which species such as bulrushes and yellow iris may grow. Except at Clydebank, the reed fringe can be found on both sides along the majority of the canal's length. The walls of the canal support several ferns (including male-fern, lady-fern and broad buckler-fern) together with more terrestrial species such as rosebay willowherb and occasionally hogweed and orchids. Meadows bordering the towpath provide nectar bearing plants for insects and habitats for some small mammals. Mammals associated with the canal itself include mink, otters (which feed in the canal and breed on the nearby River Kelvin), possibly water voles and bats in some culverts and tunnels.

Several important sites are associated with the urban stretch of the canal. One of these, Possil Marsh, is bounded by the canal and a residential area to the south and east, and a major road, a cemetery and a housing estate to the west (Plate 4.4). This 28 ha site comprises standing water around which are wetland vegetation communities such as reedbeds (including bulrushes), wet meadows and marshy areas (Lampard and Morgan, 1991). It was Scotland's first nature reserve in 1931, and was designated a SSSI in 1954 (Forth and Clyde Canal Joint Advisory Committee, 1995). The area is a Statutory Bird Sanctuary and a total of over 150 species of birds have been recorded from it, around fifty of which are known to have bred on-site. It is of major interest for those species breeding and feeding on open water and amongst reedbeds, and is an important feeding site for many migratory waterfowl and waders. In addition, several birds of prey (kestrel and sparrowhawk) and owls (long-eared and short-eared) also occur. It is not only the bird community which is diverse; many invertebrates and over 250

Plate 4.4: Possil Marsh

species of plants have been recorded from the site.

Other sites along the urban length of the canal, although less important for nature conservation, are still of local interest. One such is the Garscadden section of the main canal, which comprises a flight of locks with short pounds between. The towpath alongside is a favourite area for dog walking, strolling, etc. The towpath vegetation of scrub and trees includes large stands of Japanese knotweed, some over 25 m long. This stretch of canal is cut off from navigation; there is a 400 m culvert at the northern edge of the section beginning where the canal goes under the A82. On the non-towpath side are shallow areas with reedbeds where waterbirds breed, and the area is known for its ducks and coots, which are fed by local people.

Another heavily urbanised site is Firhill Basin, near to Partick Thistle football ground on the Glasgow Branch of the canal (Plate 4.5). Here, the canal has been widened to create two basins, separated from each other by a curved island. The bridges, which once connected the island to the shore, have been removed affording the wildlife communities living there considerable protection from disturbance. Terrestrial plant communities of tall grass and herbs are bounded by large reedbeds of sweet reed-grass and bulrush, together with willow scrub. The area also contains a range of animal life including breeding waterbirds such as mute swans.

Pressures on the urban part of the canal and its attendant habitats are mainly from local people, including damage caused by burning and dumping of rubbish from surrounding areas. There are a few very localised pollution sites, but generally the water is surprisingly clean, supporting rich populations of invertebrates. Management of vegetation includes partial cutting of towpath verges (1 m on each side) and amenity areas, and periodic clearance of duckweed to prevent reduction in light transmission. Today, further infilling and developments which interfere with navigation have stopped, large sections have been cleared of silt, weed and rubbish, and work has begun to reopen more to navigation. Plans exist to

Plate 4.5: Firhill Basin

clear further sections of canal, eventually involving the whole canal from coast to coast. However, this requires considerable finance since several stretches need to be deepened, others at present infilled must be dug out, and several low road bridges need to be dealt with. The £78 million Millennium Link Programme (which covers Scotland's Lowland Canals: Forth and Clyde and the Union Canal) has recently won support of £32 million from the Millennium Commission and work is now in progress to link the Forth to the Clyde and Glasgow to Edinburgh. Allied to these improvements for navigation is the maintenance of many important sites for aquatic wildlife and environmental education.

A STORMWATER RETENTION BASIN IN THE WEST MIDLANDS

Several designs have been proposed to control, treat and avoid the problems created by urban stormwater (see Marsalek *et al.*, 1993 for review). Examples include the use of detention or retention basins (Simmons and Barker, 1989). Detention basins are dry depressions designed to temporarily hold stormwater and slowly drain once peak levels have passed. Retention basins are areas of open water or wetland which can accept extra volume during stormwater events. The latter are better for wildlife since they provide a permanent range of aquatic habitats which are usually only swamped in exceptional circumstances. In both designs, sediment settles out while water is held in the basin. An additional benefit of retention basins is that flood events tend not to resuspend these sediments. This settling capacity, together with the action of reedbeds where present, is effective in removing pollutants (Simmons and Barker, 1989). For particulate matter, including lead, in excess of 90

per cent may be removed in this way, and high levels of more soluble pollutants such as phosphorus (65 per cent) and organic matter, nitrogen, zinc and copper (50 per cent of each) can be removed. However, the accumulation of pollutants in sediment can eventually cause problems of water quality, and in disposal of dredged sediments. Both types of basin have been built in the United States, Canada, Australia, and more recently in Britain. In addition to their primary function to control both quantity and quality of stormwater, many are managed for wildlife, especially birds. With careful design and management, waterfowl, waders and marsh-living species can all be catered for. Amongst the basins in Britain, are a series of balancing lakes in Milton Keynes (Hengeveld and de Vocht, 1980/81) and several retention basins around Birmingham, one of which, Forge Mill Lake, is the subject of this case study.

Forge Mill Lake, situated in Sandwell Valley, is owned by the local council, although 25.5 ha of the eastern part is leased to the Royal Society for the Protection of Birds (RSPB) who have built several hides and a visitor centre. The lake is intended to cope with stormwater from the north east via the River Tame. The site was originally a dry depression, occasionally flooded by the adjacent river, and was bounded by a landfill site and a colliery spoil heap. The permanent lake was formed in 1980 by digging out the depression further, supplementing the existing high ground and surrounding banks with the material removed, and building a sluice system. In the last twelve years the site has been flooded by stormwater four times, although not to capacity. In recent years the water level has been slightly reduced, possibly due to low rainfall. The River Tame, before recent management, was grossly polluted with industrial discharges, leachate from spoil heaps and treated effluent from sewage works. In recent years the water authority and local councils have improved its condition, and fish are now more common. Purification lakes (including a reed bed) upstream undoubtably help with this. Dredged shingle from the River Tame, recently used to extend an island in Forge Mill Lake, was checked for contaminants and found to be acceptable.

The 12 ha lake has a central area of deep water (about 1.5 m) with two regions of shallows (approximately 0.5 m deep), one to the east and the other to the north (Figure 4.6). Within the shallows are low islands which increase the amount of perimeter available for breeding ducks, such as mallard and tufted duck. The northern island has been recently extended, doubling its previous area and, by building two spurs, more than doubling its shoreline. The marsh to the east has been enlarged, although it is vulnerable to drying out and requires regular dredging to maintain the water level. The original intention was to

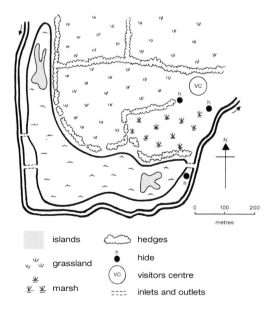

	islands		hedges
	grassland	h	hide
	marsh	vc	visitors centre
			inlets and outlets

Figure 4.6: Map of Forge Mill Lake.
Source: Lewis & Williams (1984)

utilise the deeper water for water sports, with the majority of the shore and both shallows being reserved for wildlife. In practice, little recreation takes place even in the deep water. On about seven occasions each summer, water sports normally held nearby move onto Forge Mill Lake, but otherwise the site is not disturbed by recreation. Angling is prohibited, although there is occasional illegal fishing, associated with the deliberate introduction by anglers of some fish including pike, roach, common bream, tench and a few carp. In the early days of the management of the balancing lake, problems were experienced when leachate from the landfill site increased eutrophication of the lake. In order to halt the subsequent weed growth, chemical control was employed. Although low concentrations of herbicide were used, previously undiscovered underwater springs enhanced water circulation. As a consequence, too much weed was killed too quickly, and waterfowl numbers fell. However, this problem has been resolved and many waterbirds now occur on the lake and surrounding area, including mute swans, Canada geese, wigeon, teal, shoveler, pochard, tufted ducks, mallards, coots, snipe and common sandpipers (Welch, 1986; Hextell and Hackett, 1994). Several, including tufted ducks and wigeon, have increased in the locality since the construction of the balancing lake. Casual sightings on the lake include garganey, common scoter, gadwall, scaup, ringed plovers, greenshank and green sandpipers.

To the north of the lake, and bounded by the River Tame to the west, is a wet meadow which is planned to be managed as a water meadow, grazing cattle in the summer and providing a habitat for ducks such as wigeon in winter. The River Tame runs around the south and west sides of the lake and supplies the water, especially when in flood. The surrounding grassland (some of which is subject to winter flooding) provides feeding grounds for invertebrates and birds including the ubiquitous Canada geese. On the river banks are several established trees and on the higher ground to the east are small clumps of trees planted about fifteen years ago. Further tree planting is planned around the lake, including stands of willows, which will reduce the wet grassland available to Canada geese (which are something of a pest), provide more habitats for warblers, and enable coppicing for craft work. A double line of old hedges runs from the eastern bank, which also increases the habitat diversity for a range of wildlife.

Sandwell Valley as a whole is managed for recreation, education and wildlife and the range of habitats at Forge Mill Lake are replicated, to a greater or lesser degree, at several other ponds and lakes nearby. This is important, as it provides a network of habitats which can act as refugia for animals if disturbance is problematic at any one site. The diversity of habitats (lake, river, marsh, woodland, grassland, scrub and hedgerow) not only provides sites for many bird species, but also for mammals, invertebrates and plants. The range of terrestrial habitats in the Sandwell Valley area also add to the diversity, with many species of plants (293), fungi (180), birds (167), insects (1408, including 284 beetles, 279 moths, 19 butterflies, 115 hoverflies), amphibians (4) and mammals (17) having been recorded (Bloxham, 1986).

COASTAL HABITATS IN CARDIFF

The coastal region of Cardiff is important for nature conservation although major parts are under threat from developments. Three rivers enter the Bristol Channel having run through Cardiff: the Taff and Ely estuaries converge at Cardiff Bay and the Rhymney emerges about 5

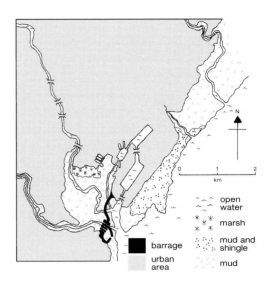

Figure 4.7: Map of the coastal area around Cardiff.
Source: City of Cardiff (1995)

km to the north-east (Figure 4.7). The estuarine mudflats and saltmarshes at the mouths of these estuaries are very important habitats, especially for waders and waterbirds. These estuarine sites comprise two of the fourteen SSSIs associated with the Cardiff area (three of which are geological SSSIs) and, with the exception of the nearby island of Flat Holm, are the only marine SSSIs in Cardiff. All three rivers are major wildlife corridors and are associated with several Sites of Nature Conservation Importance and two local nature reserves. Inland, important areas include reedbeds, marshland, wet meadow, rough grassland and ancient woodland. At the coast, the tidal river, mudflats and saltmarsh are the most ecologically important, providing habitats for migratory fish, otters, wildfowl and a range of bankside vegetation and associated fauna. Following a reduction in coal extraction in the region and the subsequent cleaning of the rivers, migratory fish such as salmon and eels have moved upstream over recent years.

Both major areas, the Rhymney Estuary and Cardiff Bay, are threatened by urbanisation in the near future. A planned extension to a landfill site near the mouth of the Rhymney will spread into an area designated as a SSSI on account of the flora and fauna which inhabit the drainage system on the site, known as reens and ditches. Mitigation for this plan involves the development of new sites near to the extension. The mudflats and saltmarshes at the estuary itself should be undisturbed by the development. This is very important since they form part of the Severn Estuary SSSI, a fairly continuous coastal belt which, together with other habitats along the Severn, is one of the largest and most important intertidal areas in Britain (Marine Task Force, 1994). The Severn Estuary is particularly important for fish (over 110 species have been recorded) and birds (especially waders and wildfowl which reach peak winter numbers of over 50,000 and 25,000 respectively).

The port of Cardiff is still commercially active, albeit at a much lower level than in its peak during the early 1900s (Beresford, 1995). As part of a major redevelopment programme of the dockland area, a 200 ha freshwater inland bay for recreation and commerce will be created by building a barrage across the mouth of Cardiff Bay to separate the outflow of the Rivers Ely and Taff from the Bristol Channel. Extensive mudflats at the mouths of the Ely and Taff are the result of deposition from the rivers and land reclamation programmes in the bay during the last century. Currently, these are important winter feeding sites for around 8,000 birds and the site has one of the highest densities of birds in the Severn Estuary (City of Cardiff, 1995).

The barrage has been the subject of major debate. The arguments in favour of the development mainly reflect economic and recreational aspects, including the viewpoint that the mudflats restrict the use of the bay and are unsightly. The tidal range is high (up to 14 m) and currently much of the bay is inaccessible from the sea for over 12 hours per day. One anti-barrage argument has concentrated on the impacts on the visiting birds that currently feed in large numbers on the mudflats. There are additional worries that an increase in recreation in one part of the Severn Estuary will encourage expansion elsewhere, to the detriment of wildlife (Salmon and Fox, 1994). Supporters of the barrage argue that Cardiff Bay is a relatively small part of the Severn Estuary SSSIs, giving displaced birds alternate feeding grounds close by. In addition, new areas for wildlife are being developed by the Cardiff Bay Development Corporation in association with the Countryside Council for Wales and the Royal Society for the Protection of Birds. However, there is no doubt that the impact on birds will be significant, especially bearing in mind planned developments of other barrages across the Rivers Severn, Mersey, etc., which will further restrict winter feeding grounds. It has been estimated that the effects of the proposed Severn Barrage on the intertidal zone could remove feeding grounds for around 23 per cent of current usage of ringed plover, 39 per cent of shelduck, 50 per cent of dunlin and 80 per cent of redshank (Clark and Prys-Jones, 1994).

Another anti-barrage argument is that sewage and other effluents (which currently flow down the Ely and Taff into the Bristol Channel) will build up in the enclosed bay. However, the existing sewage system will be uprated to reduce the amount discharged (Jones, 1994). Other water quality issues include risk of deoxygenisation (a problem with deep still waters) and algal blooms (some of which produce toxins). Oxygen levels will be maintained by mixing and oxygen injection, and excessive algal growth will be removed. Provision will also be included to drain the bay in the case of a major pollution incident. Much of the barrage is now in place across Cardiff Bay, with final impoundment due late in 1998. The 800 m embankment projecting from the north of the bay will be linked to the southern edge by five sluice gates and three locks. The sluices allow the water from the two rivers to exit the bay, while the locks permit the passage of boats. Potential problems associated with access by migratory fish to the river network will be accommodated by the incorporation of a fish pass. It will be important to monitor the barrage development to assess whether the feared impacts on wildlife do occur, and to examine the success of the new wildlife sites being developed in mitigation.

5

PRACTICAL WORK

•

By far the best way of understanding the ecology of a habitat is to investigate it for yourself. In order to gain maximum benefit from ecological studies it is important to plan in advance. Part of the planning is to be sure of the questions being asked or problems to be solved; the most elegant research tends to ask simple questions or look at simple problems. Prior planning also includes identifying the methods for data analysis (see below). Equipment should be checked and the experimenter should be fully competent in its use before gathering data. A consistent approach, using the same techniques and only varying the factor to be analysed is usually the best approach. Often a small pilot study will help the main investigation to run smoothly and will allow the methods to be refined.

It is important that permission is received from the owner(s) of land being used for the project, both to gain access and to carry out particular types of experiment. For example, it is against the law (Wildlife and Countryside Act: Department of the Environment, 1981) for any unauthorised person (i.e. anyone who is not the owner, occupier or has been authorised by the owner or occupier of the land concerned) to uproot any wild plant. Other organisms are even more strictly protected; almost all birds and some other animals, together with several plant species, are fully protected under the law (see Jones, 1991, for further details). Care should be taken to comply with the law, to ensure that you do not

damage or unduly disturb any plant or animal, and to design and implement any research project so as to leave the habitat as it was found.

Urban habitats provide a wealth of topics which may be investigated throughout the year with relatively little specialised equipment. Urban systems, including urban–rural gradients, provide opportunities for ecological research into soil and water resources, air, water and soil pollution, population and community characteristics and ecosystem structure and function (McDonnell and Pickett, 1990). The following projects may be investigated with relative ease, require little in the way of equipment and utilise common urban habitats or species. They are designed to ask simple questions and provide data to analyse with simple tests. As in most research, the results obtained may stimulate further work on supplementary questions.

EXPERIMENTAL DESIGN

Comparisons between situations in which only one factor varies are the easiest to interpret. For example, comparisons between several sites which differ in size, but are similar in all other respects will allow an examination of the influence of size of site on whatever is being recorded. The experimental design must be considered in some detail. There are two major types of experimental design: observational and manipulative. The majority of projects described below are observational. Here, a variable (the behaviour of animals, the percentage cover of different plant species, the numbers of animal of a particular species, etc.) is recorded under different circumstances (different sites, weather conditions, numbers of animals in a group). Analysis is often a matter of finding whether the variable measured differs in two or more circumstances (e.g. sites, times of day, etc.). If two variables (e.g. number of species and temperature) are measured, you may wish to examine the relationship (called correlation) between the two. Two variables may have a positive (as one increases, so does the other), negative (as one increases, the other decreases) or no relationship. However, with observational experiments you will not be able to say definitively that a change in one variable causes the change in the other. This is because an observed increase in one variable (e.g. the number of a species) may be correlated with a measured rise in another factor (e.g. temperature) but actually be due to changes in a third, unmeasured variable (e.g. the abundance of its prey). In manipulative experiments, on the other hand, the experiment is designed such that one variable is altered by the experimenter (pH, temperature, fertiliser concentration) allowing a much greater emphasis on cause and effect when it comes to analysing the data. However, since these are often conducted in fairly artificial conditions, they may have less relevance to real world situations than may be achieved with observational experiments.

It is important to take several replicate samples, since data gathered from only a single sample may not be representative of the situation in general. In order to reduce bias, sampling should be systematic or random: instructions on sampling are given in each practical. It is essential that data gathered is recorded correctly to avoid later problems in analysis and interpretation. In order to help you collect data in an appropriate way, sample data recording sheets have been included where appropriate. Record your data in a hard-backed book rather than loose sheets of paper to reduce the risk of loss. Where data are to be analysed using a computer, it is useful to transcribe the data onto an appropriate spreadsheet or other data file as soon as possible, and to check the transcription carefully to identify and rectify any mistakes. The techniques which could be used in the analysis of such data are beyond the scope of this book, although methods of analysis are suggested in the project descriptions, and several texts which provide further details are listed in the further reading section.

HEALTH AND SAFETY

Health and safety in field (and laboratory work) should be paramount. Do not engage in behaviour or activities which could harm yourself or others, and assess the risks and health and safety issues which are likely to be involved and protect against them. General safety issues include: wearing appropriate clothing for the time of year and terrain; using safe equipment (e.g. plastic tubes rather than glass); not working far from help (ideally work in groups of two or more); informing someone responsible of the details of planned field work in advance (including location and duration) and 'signing off' with that person on return.

Soil and water contain organisms and compounds which are hazardous to health. Gloves should be worn for all fieldwork involving these materials, and when handling spiny, or otherwise hazardous plants. Tetanus is a potential hazard for anyone working out of doors, especially those in contact with soil.

Spores of the tetanus bacillus live in soil, and minor scratches (e.g. from bramble thorns) could provide a point of infection. Immunisation is the only safe protection and, being readily available, should be kept up to date. Weil's disease (or Leptospirosis), is caused by a bacteria carried by rodents (especially rats). Urine from infected animals contaminates freshwater and associated damp habitats such as river, stream and canal banks, and is more common in stagnant conditions and during warmer months. Infection is usually via cuts or grazes, or through the nose, mouth or eye membranes, and precautions should be taken to avoid contact between these areas and potentially infected water. Cover cuts and grazes with waterproof dressings, use appropriate waterproof clothing including strong gloves and footwear, and avoid eating, drinking or smoking near possible sources of infection. Lyme disease is another potential hazard for fieldworkers. This is transmitted by female ticks which, although more usually a problem in uplands and woodlands, have been found in urban parks. They are especially likely to bite from early spring to late summer. To help avoid the disease, prevent ticks from biting by wearing appropriate clothing (e.g. long trousers), check for the presence of ticks (light coloured clothing helps) and remove ticks as soon as possible if bitten (twist slowly in an anti-clockwise direction without pulling and seek medical help if mouthparts remain within the skin).

In addition, it must be remembered that many urban habitats are potentially dangerous places; watch out for obvious hazards such as sharp pieces of metal, plastic and broken glass, avoid traffic hazards when working on streets and roadsides and be careful on uneven surfaces. The safety issues indicated here are not fully comprehensive. Before any fieldwork is undertaken you are advised to consult appropriate publications such as that produced by the Institute of Biology (Nichols, 1990).

DISTRIBUTION OF PAVEMENT PLANTS

Pavements are ideal habitats in which to examine the impacts of some stresses on plants (see Chapter 2). This project investigates the relationships between vehicle and pedestrian density and the plants growing in cracks in pavements. Select six to ten streets with similar widths of pavement and sizes of paving stones, but which differ in the levels of traffic using them. For example, you might choose the following: a main road; a minor road; a large residential street connecting two main roads; a large residential street with no such connections; a small residential avenue; a cul-de-sac. It is best not to use city-centre streets, since they are often devoid of plants and the large number of pedestrians makes them hard to survey. Measure a 25 m continuous strip of pavement along one side of the street, starting a set distance (say 10 m) from the beginning of the street or from a major junction with a different type of street. Within this strip, record (see Table 5.1) how many vehicles and pedestrians pass on that side of the street in 15 minutes. Sample the plants growing in the gaps running parallel to the kerb, both sides of the paving stone nearest the kerb stone (i.e. 50 m of total gap or 2×25 m: see Figure 5.1). Ignore gaps running at right angles to the kerb, in order to standardise the length of habitat surveyed, because kerb stones of different sizes would produce differing total gap lengths. Count the total number of plants (if individuals are difficult to distinguish, then use the number of clumps, but be consistent) and the number of plants of each growth type shown in Figure 5.2. Repeat this for the other side of the street and combine the results for each

Table 5.1:   Recording sheet for the pavement plant project

Date..									
Street	No. of vehicles in 30 minutes	No. of people in 30 minutes	Total number of plants / clumps per 100 m	Number of mat-like plants per 100 m	Number of grasses per 100 m	Number of rosette plants per 100 m	Number of tall herbs per 100 m	Etc.
Etc.								

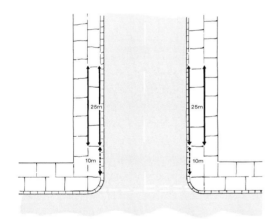

Figure 5.1: Pavement area to be sampled.

street (by adding the values you now have data for plants from 100 m of pavement gaps and for 30 minutes of traffic and pedestrians). Using both sides of the street in this way takes into account differences due to aspect; for example, south-facing pavements are exposed to different conditions than north-facing ones.

This project enables the identification of relationships (using rank correlation coefficients) between the total number of pavement plants or the number of each growth type and:

(a) traffic density;
(b) pedestrian traffic density.

Bear in mind that the levels of traffic and pedestrians recorded are merely a snapshot in time. You need to consider how representative your 30-minute survey is of typical levels. It is not essential to identify each species of plant for this particular project, although the project can be extended by examining how many different species there are. In addition, if species are identified, then supplementary questions can be asked, including examining the numbers and mean distances from the kerb of common species in relation to vehicle and pedestrian density. Looking up species characteristics in plant ecology texts (e.g. Grime *et al.*, 1988) will enable interpretation of these data with respect to the ecology of the species concerned.

FLORA OF URBAN WASTELAND

The colonisation of wasteland by plants has been documented earlier (Chapter 2; see also

Figure 5.2: Common pavement plants.

Erect plants;   (a) rosebay willowherb; (d) prickly sowthistle; (e) smooth sowthistle.

Rosette plants   (b) dandelion; (c) greater plaintain; (i) shepherd's purse.

Grasses: (g) annual meadow grass; (h) perennial rye grass.

Low growing plants: (f) mosses such as silvery-thread moss; (j) procumbent pearlwort.

Gilbert 1989 and 1992a). Many of these plants are clumped in distribution, often deriving from individuals which establish in pockets of suitable conditions (Colour Plate 8). Their life history (whether annual, biennial or perennial), their reproductive strategies (sexual or vegetative) and dispersal mechanisms (by wind, in soil or on animals) vary between species and affect the likelihood of colonisation and distribution. The aim of this project is to investigate the distribution of plants in urban wasteland and to decide whether they are randomly distributed, clumped or evenly spaced. The distance that each plant is from the nearest neighbouring plant of the same species can be compared to the theoretical distance which would be expected if the distribution was random. If the mean observed and expected distances differ widely, then we assume the plants are not randomly distributed. First carry out a preliminary survey using random coordinates to select points at random. To do this you need to lay out a grid using two tape measures at right-angles to each other along the sides of your site to form two axis. Using pairs of random numbers (obtainable from a calculator or book of statistical tables) you can identify a position relative to each axis. So if your axes are 100 m long, and your first pair of random numbers are 10 and 63, then your random point is 10 m in one direction and 63 m in the other. Where

an axis is of a different length to 100 m, you will need to divide the axis length by 100 before multiplying by the random number: so an axis length of 25 m and random numbers of 20 and 35 will result in a 5 m (0.25×20) and 8.75 m (0.35×35) coordinates.

Place ten 2×2 m quadrats at random coordinates on an area of wasteland, and within each count the number of plants of each species. From these data, ascertain which five species are the most frequent; their identification can be confirmed using a field guide (see further reading section) or flora (e.g. Stace, 1997). From the quadrat data, calculate the density (in plants per square metre) of each species. For example, if you found 8 rosebay willowherb plants in your ten random quadrats, then the density is 8 divided by 40 (because the area sampled is 10 quadrats each 2×2 m in size) or 0.2 plants per m².

Next, use a new set of random coordinates to locate twenty points on the wasteland. At each point find the nearest plant of each of the five most common species. For each plant of each species record (Table 5.2) its maximum width and the distance from the centre of the plant to the nearest neighbour of the same species (use metres for all your measurements). Use the formulae below to calculate a single value for each species which represents its distribution.

$$d = \frac{(Observed\ mean\ distance\ between\ the\ plants - Expected\ distance\ between\ the\ plants)}{Standard\ error\ of\ the\ observed\ mean\ distance\ between\ the\ plants}$$

Where:

$$Expected\ distance\ between\ the\ plants\ if\ distributed\ randomly = \frac{1}{\sqrt{(2 \times Density\ of\ the\ plants)}}$$

and:

$$Standard\ error = \frac{0.26316}{\sqrt{(number\ of\ plants \times density\ of\ plants)}}$$

Table 5.2:   Recording sheet for the wasteland flora project

Date							
Point number	Nearest individual of species A		Nearest individual of species B		Nearest individual of species C		Etc.
Point	Maximum width (m)	Distance to nearest neighbour (m)	Maximum width (m)	Distance to nearest neighbour (m)	Maximum width (m)	Distance to nearest neighbour (m)	
1							
2							
Etc.							

The lengths must be measured in metres and the density calculated as the number of plants per square metre. If d is greater than 1.96 then the plants are significantly evenly distributed, if d is less than −1.96 then the plants are significantly clumped. Between these values we assume that the plants are randomly distributed. This project will allow you to examine the distribution of common plants of wasteland with respect to:

(a) characteristics of the species concerned (the life history, reproductive and dispersal characteristics can be looked up in an appropriate text, e.g. Grime *et al.*, 1988; or Stace, 1997);

(b) the size of the plants by examining the relationship between the width of the plant and the distance to its nearest neighbour using rank correlation coefficients.

You could extend the project by looking at several sites which differ in the time since they were abandoned, identifying where they could be placed in the successional series described by Gilbert (1989 and 1992a; Table 2.2), and deciding whether this agreed with the expected stage according to the time since abandonment.

DISTRIBUTION OF TAR-SPOT FUNGUS ON SYCAMORE TREES

Sycamore leaves often have distinctive black patches caused by the tar-spot fungus (Plate 5.1). The fungus overwinters on fallen leaves, maturing to produce spores in the following spring and early summer. The spores are released into the air and borne by wind to infect new leaves. After 2–4 weeks, a black spot appears (1–2 mm in diameter) which increases to about 7 mm over the season. A 1–2 mm band of leaf around the spot is usually chlorotic (lacks chlorophyll and appears yellow). The fungus attacks many species of *Acer*, but is especially frequent on sycamore. It seems to cause the tree little damage. Until recently, the fungus was less common in cities than rural areas, leading to suggestions that it

Plate 5.1: Sycamore leaf infected with tar-spot fungus

was intolerant to high sulphur dioxide levels (Edlin and Mitchell, 1985; Cannon and Minter, 1984). However, a study in Edinburgh indicated that pollution levels are less important than the distance from infective sources of the fungus, and that the observed reduction of infection in towns may be due to removal of overwintering leaf litter by street cleaning (Leith and Fowler, 1988). This leaves an interesting problem. It is possible that the fungus does require local sources of infection. However, the relationship with sulphur dioxide may not be simple. There may still be an intolerance to sulphur dioxide at levels which are not found except in larger cites (Edinburgh, in common with several other small cities, has levels which are not a great deal higher than in some rural areas: Bower *et al.*, 1995). In addition, air quality may have previously been important but recent reduced levels may have fallen below those which affect the fungus.

The aim of this study is to examine the distribution of tar-spot fungus on sycamore with respect to factors which may be involved in reinfection. Simple comparisons between infection levels on trees in clumps with those which stand on their own may help to discover whether proximity of infection source is important. However, there is a possible complication: isolated trees are more exposed to pollution than are those in a clump, therefore any reduction in infection in isolated trees could be due to pollution rather than distance from a source. We can investigate this possibility by comparing trees from the outside of clumps (more exposed) to those within (less exposed). In mid to late summer, select several study sites which have isolated sycamore trees and others with clumps of at least ten sycamore trees. Use trees of as similar size as possible in each case and have as many sites of the two types as you can manage (about ten of each will probably suffice). Collect ten leaves at random from each isolated tree and from a tree near the centre of each clump. Using random-number tables, each number obtained (e.g. 11, 23, 56) will enable a random selection of leaves to be made (e.g. the 11th, 23rd, 56th leaf encountered). Make sure that you collect all leaves from about the same height on each tree. Count the number of fungus spots on each leaf and calculate the mean number of spots per leaf for each tree (Table 5.3). Measure the distance (in metres) to the nearest source of fallen sycamore leaves (i.e. the nearest sycamore tree), recording 100 if there is no source within 100 m of the tree. In addition, randomly select a tree on the outside of each of your ten clumps and collect ten leaves from the side facing away from the clump, and another ten from the side facing into the clump. Calculate the mean number of spots on leaves on the exposed side and sheltered side of each tree.

This project enables the incidence of infection (measured as the mean number of tar-spots per leaf for each tree) to be examined in relation to:

(a) the two types of tree group (individual and centre tree of the clump);
(b) the distance from potential sources, analysed separately for the individual trees and those in the centre of clumps;
(c) whether leaves are sheltered or exposed (possibly to pollutants) by comparing leaves from the sheltered and exposed sides of trees on the outside of clumps;

(a) can be analysed with t-tests or Mann-Whitney U tests, (b) using rank correlation coefficients, and (c) with paired t-tests or Wilcoxon's test for matched pairs. The results may help to elucidate whether leaves on trees close together and isolated trees in the same neighbourhood have differing numbers of tar-spots, and if distance to a source of infection and/or exposure are related. This project could be repeated in a rural area and the results compared to those from an urban area. Remember that the conditions at each site should be as similar as possible to each other.

COLONISATION OF STONES BY INVERTEBRATE ANIMALS

Stones and rubble resting on the ground provide a habitat for an interesting mixture of surface-active and soil animals. These include predators (such as ground beetles, spiders, harvestmen and centipedes) as well as herbivores (including slugs and snails) and decomposers (earthworms, woodlice, millipedes and springtails). Such animals are temporary visitors using the sites for part of the life-cycle (e.g. mating aggregations, egg-laying,

Table 5.3:   Recording sheet for the tar-spot fungus project

Date				
For individual trees and those in the centre of clumps			*For trees at the edge of the clumps only*	
Group size (*I = individual; C = clump*)	Mean number of tar-spots per leaf	Distance of tree from nearest source of infection	Mean number of tar-spots per leaf on the exposed side of the tree	Mean number of tar-spots per leaf on the sheltered side of the tree
Etc.				

hibernation or aestivation). Some animals use stones as refuges from extremes of micro-climate (drying or freezing) or to avoid preda-tion. Others are searching for food themselves in the form of decaying plant material, fungi and other animals.

The communities under stones may be sam-pled by hand. If the stone is carefully lifted then, with a little practice, the animals may be picked up, preferably using a pair of soft-bladed forceps, and placed in a plastic tube. Small animals can be collected using a pooter (see Figure 5.3). Be careful to keep animals moist by placing small pieces of vegetation with them and keep predators in separate con-tainers to avoid them attacking other speci-mens. Some animals freeze when disturbed and are difficult to see while others dart away swiftly; you will need a good eye and quick hand to catch the whole community. Before beginning the project practise your technique until you are quite competent, otherwise you may find that you under-record from the first few stones you sample.

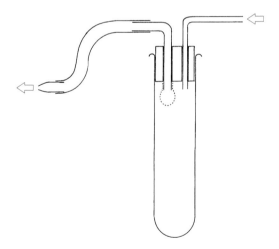

Figure 5.3: Pooter comprising a tube with two pieces of tubing leading from it, one has a larger bore and is sucked through while the other piece is held over the animal which will be sucked into the collecting tube.

The dynamics of colonisation and estab-lishment can be investigated by placing stones or similar items on the ground, observing them over a period of time and recording the rate at which different animals colonise underneath. It is important that the experiment takes place in a homogeneous environment, where the stones will remain undisturbed for the sam-pling period. Select stones of a similar size, shape and material; not too large to lift easily, or too small to provide protection for animals. Flatter stones tend to work best. The number used depends on the size of the area (they should not be too close) and the time avail-able; twenty stones, up to paving stone size will probably be sufficient. These should be laid on the ground in a grid pattern at least 1 m apart. Give each stone a unique number using waterproof marker pen or paint.

Spring and autumn are probably the best times to carry out this project, since in severe winter and summer weather soil animals tend to burrow deeper, and those which are usually surface-active hide in crevices. Compare col-onisation after one week with that after two weeks. Leave the stones *in situ* for a week, then sample half the stones by carefully lifting alternate stones, one at a time, and collecting the animals. Identify the animals and release them elsewhere on the site away from the experimental area. After a further week exam-ine the remaining stones and collect their resi-dent fauna. Record the data as shown in Table 5.4 so that each stone is identified by its unique number, the type of collection is dis-tinguished (i.e. whether collected after one or two weeks), the total numbers of animals, number of groups (i.e. earthworms, slugs, snails, etc.) and the numbers of animals in each group are noted.

Animals can be identified using the informa-tion in Table 5.5 and Figure 5.4, or specialised texts (e.g. Wheater and Read, 1996). A hand

Table 5.4:   Recording sheet for the colonisation of stones project

| Site details... | | | | | | | |
Stone number	Collection (collected after one week, or two weeks)	Number of animals	Number of groups	Number of earthworms	Number of slugs	Number of snails	Etc.
Etc.							

lens (×10 or ×20) is useful. The animals can be classified according to their ecological role (predatory, herbivorous, soil active, surface active: see Figure 5.4 and Table 5.5). Compare the numbers of animals, groups (as shown in Table 5.5), animals in each group, predators, herbivores, decomposers, soil animals, and surface-active animals between each of the periods of sampling (one week versus two weeks) using t-tests or Mann-Whitney U tests.

It may be useful to record any changes which take place in the conditions of the experimental site over the recording time. For example, if the weather changes appreciatively, or if the stones were disturbed during the sampling period. This project could be extended by looking at longer time periods, or by examining stones which differ in size or material.

DISTRIBUTION OF GREY SQUIRREL DREYS

Grey squirrels are common in urban areas, especially in parks. They feed mainly on tree seeds, fruit and buds (Gurnell, 1987), especially from hazel, oak, field maple, ash, beech and hornbeam (see Species Box 2.3). The presence of their dreys in trees is a useful measure of the number of grey squirrels present, particularly in winter when dreys are most obvi-

ous (Don, 1985). Dreys are large domed nests, usually built 5 m above the ground, with an outer layer of twigs and leaves, and an inner layer of softer bedding material (Plate 5.2). They are fairly obvious and are frequently situated in the forks of branches, often near to the main trunk. Some bird's nests (e.g. magpies) may look somewhat similar, but tend to be higher in the canopy and often of a looser construction. In a study of farm woodlands, Fitzgibbon (1993) found that occurrence of oak, beech or hazel increases the likelihood of dreys being present. The aim of this project is to identify whether grey squirrels in urban parks are associated with certain tree species.

Count the number of trees of each of the five most common species which are appropriate for drey building (i.e. several must be at least 5 m tall) and count the number of each tree species in which there are one or more dreys and the number in which there are no dreys (see Table 5.6). Preferably record only those dreys which are still being used; disused dreys tend to be easier to see through once they start to deteriorate. Count the number of major branches above 5 m from the ground and calculate the mean number for each species. Estimate the height of each tree examined and calculate the mean height for each species. Heights of trees, or any other tall object, can

Table 5.5:   Common animals under stones

Make sure that your specimen matches all parts of the description. Note that these are not the only animals which may be found (for further information see Wheater and Read, 1996).

Animal group	Description	Information
Earthworms	Body composed of more than 20 segments; no legs	Soil animals feeding on plants, fungi and decomposing material
Slugs	Body without segments; no legs; head with tentacles	Surface-active animals feeding on plants, fungi and decomposing material; a few are predatory
Snails	Body without segments; no legs; with external shell; head with tentacles	Surface-active animals feeding on plants, fungi and decomposing material; a few are predatory
Harvestmen	Body composed of one distinct region with eyes sitting on top; four pairs of legs, of which the second pair is the longest	Mainly surface-active predators, occasionally feeding on dead invertebrates, bird droppings, fungi and plant matter
Spiders	Body divided into two distinct regions with eyes at the front; four pairs of legs	Entirely predatory, some use webs while others actively hunt their prey on the surface
Pseudoscorpions	Body divided into two distinct regions; set of pincers at the front; four pairs of legs	All are predatory, using 'pincers' (pedipalps) to catch prey within the leaf litter
Mites	Body composed of one distinct region; two or four pairs of legs, second pair not the longest	Many are parasitic; free-living species are predatory or feed on fungus and decaying matter in the soil and leaf litter
Woodlice	Body less than three times as long as wide; seven pairs of legs, with no more than one pair per segment	Mainly in and on leaf litter feeding on dead and decaying plant, fungal and very occasionally living plant material
Millipedes	Body at least three times as long as wide; more than seven pairs of legs, with each segment containing two pairs	Live in leaf litter and upper soil layers, feeding on decaying leaves and dead wood; snake and flat-backed millipedes are longer and thinner than pill millipedes
Centipedes	Body at least three times as long as wide; more than seven pairs of legs, with no more than one pair per segment	The long, pale geophilomorphs are more often soil dwellers than the stouter, browner lithobiomorphs; all are predators
Three-pronged bristletails	Body divided into three distinct regions; three pairs of legs; antennae long and hairlike; three tails at end	Feed on humus and other decaying organic material in the leaf litter layer; infrequent under stones
Two-pronged bristletails	Body divided into three distinct regions; three pairs of legs; antennae long and hairlike; two tails at end	Small white insects, usually found at low densities, which feed on decaying plant material in the upper soil layers

Table 5.5: continued

Animal group	Description	Information
Proturans	Body divided into three distinct regions; three pairs of legs, front pair pointing forwards; antennae small	Small, pale and easily missed insects living in the soil and leaf litter which feed on decaying organic matter
Springtails	Body divided into three distinct regions; three pairs of legs; abdomen with at most 6 segments	Small and pale feeding on decomposing material and fungi in the soil and leaf litter, many jump (apparently at random) when disturbed
Earwigs	Body divided into three distinct regions; three pairs of legs; wings hidden under hard wing cases; pincers on hind end	Active on soil surface, they feed on a wide range of food including carrion, small invertebrates and some living plant material (e.g. petals)
Two-winged flies (larvae)	Body composed of fewer than 20 segments; no legs	Most feed on decomposing material in the soil, although some are predatory and others feed on plant roots
Ants	Body divided into three distinct regions, distinct waist between thorax and abdomen; three pairs of legs; antennae elbowed	May be found foraging on the surface for food (wide variety including vegetable matter and other small invertebrates)
Beetles (adults)	Body divided into three distinct regions; three pairs of legs; wings hidden under hard wing cases; no pincers on hind end	Ground and rove beetles are the commonest; both are active on soil surface; mainly predatory, though some eat fungi and vegetable matter
Beetles (larvae)	Body composed of fewer than 20 segments; three pairs of legs; antennae small; front legs point to sides	Mainly active in the soil with similar feeding habits to adults (though larvae of some, such as click beetles, feed on plant material)

be estimated using a clinometer (see Figure 5.5).

Frequency analysis can be used to test whether the tree species influences the occurrence of dreys (comparing the number of trees of each species with and without dreys). Interpretation of your results could be based on:

(a) the mean heights and architecture (amount of branching) of the tree species concerned (whether all species were equally suitable for building dreys);

(b) the influence of food sources (whether trees with dreys are also used as food).

The amount of available nesting sites in an area may also influence squirrel numbers (Fitzgibbon, 1993). Therefore the project could be extended by comparing several parks to examine the effect of the number of trees of a favoured species on the number of dreys. It is important to compare otherwise similar areas, i.e. the species composition of the woodlands should be as close as possible between parks.

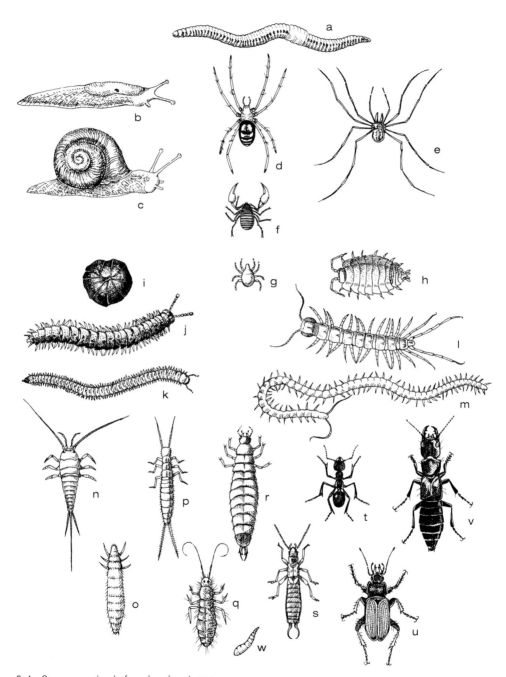

Figure 5.4: Common animals found under stones.

(a) earthworm; (b) slug; (c) snail; (d) spider; (e) harvestman; (f) pseudoscorpion; (g) mite; (h) woodlouse; (i) pill millipede; (j) flat-backed millipede; (k) snake millipede; (l) lithobiomorph centipede; (m) geophilomorph centipede; (n) three-tailed bristletail; (o) proturan; (p) two-tailed bristletail; (q) springtail; (r) ground beetle larva; (s) earwig; (t) ant; (u) ground beetle; (v) rove beetle; (w) fly larva.

Plate 5.2: Squirrel dreys

Table 5.6:   Recording sheet for the squirrel drey project

Site details .... Date				
Species of tree	Number of trees with one or more dreys	Number of trees without dreys	Mean number of major branches above 5 m	Mean height of trees
Etc.				

VIGILANCE BEHAVIOUR IN MALLARDS

There are many ways in which an animal's behaviour interacts with that of other animals and with their environment. They search for food, manipulate and consume it, protect it from others of the same and different species; they look for mates, feed and protect young; they groom themselves and watch for predators. These behaviours change depending on the circumstances (for example, during the breeding season they concentrate on feeding and protecting the young) and whether they live in groups or are solitary. Animals which live in groups may increase some behaviours (such as guarding food from others) and reduce others (such as vigilance against predation) depending on the size of the group. Reduction in vigilance occurs because of the added security of several animals watching for predators (see Krebs and Davies, 1993). The levels of vigilance by an individual may depend on the number of animals in the group and its location within the group. Those on the

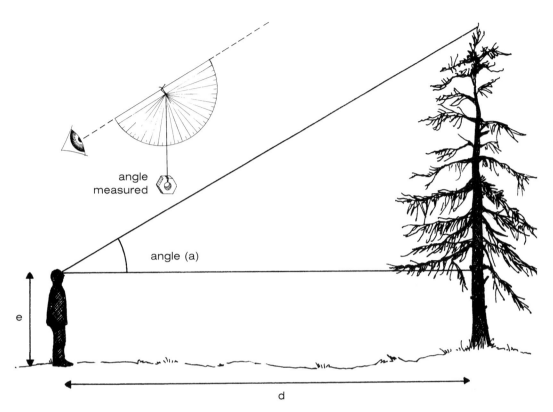

Figure 5.5: A simple clinometer can be made from a protractor which has a small hole drilled in the centre, through which is threaded a piece of fishing line with a weight on the end. Stand on the same level as the tree and about 10m away (measure this distance: d). With the flat side of the protractor uppermost and allowing the weight to hang freely, sight along the protractor to the top of the tree and measure the angle along which the weighted line hangs. The angle to the top of the tree (a) is calculated as 90 minus this angle. Measure the distance from your eye level to the ground (e). The height of the tree can be calculated using the following formula (making sure that all your measurements are in metres):

Height of the tree = (the distance from the tree × the tangent of the angle) + your eye height

outside edge may be more vulnerable than those in the centre and may therefore be more vigilant.

Mallards are common birds of urban open water and are often found in large flocks, especially where they are fed by the public. They are dabbling ducks, and spend around a quarter of their time feeding (Jorde *et al.*, 1984), usually on seeds and invertebrates sieved from the water at depths of up to 0.4 m.

They also spend time preening and watching for predators such as cats, or disturbance from humans. Differences in behaviour can be examined in relation to the size of the group and the sex ratio. On several days, as close as possible to each other, record the number of male and female mallards (see Species Box 2.24 for help with identification and sexing) on the surface of a boating lake or similar habitat (Table 5.7). Record the number of

Table 5.7:   Recording sheet for the vigilance in mallards project

Site details										
Sampling occasion	Group size	Sex ratio	Males				Females			
			Number present	Number feeding	Number preening	Number observing	Number present	Number feeding	Number preening	Number observing
Etc.										

birds of each sex which are feeding, preening, or observing their surroundings. This should be repeated on a minimum of six (preferably ten) occasions.

For each of the recorded behaviours, detect whether there are:

(a) differences between the numbers of males and females engaging in the behaviour (examined using t-tests or Mann-Whitney U tests);
(b) relationships between group size or sex ratio and the number of individuals, males and females showing each behaviour (analysed using rank correlation coefficients).

The project could be extended to look at the impacts of people on the level of vigilance by recording the numbers of visitors within 5–10 m of the flock. Alternatively you could record aspects of breeding behaviour (e.g. matings, courtship attempts), in which case the best time for the project is March, just before females lay their eggs.

GLOSSARY

•

Accretion   Growth by addition of extra material, often used in the context of siltation of rivers and estuaries.

Algal bloom   Rapid growth of algae (often in eutrophic waters) which become visible to the naked eye at high densities.

Autotrophic   Using simple compounds (usually carbon dioxide) as a source of carbon (e.g. in green plants).

Bioaccumulation   Increasing concentration of compounds (often toxins) in organisms at successive trophic levels, sometimes leading to lethal amounts at higher levels.

Canalised   Conversion of a river or stream into an artificial waterway, often involving straightening of the course and the use of artificial materials in the banks.

Cess   Cinder ballast comprising the track-bed of a railway line.

Clonal plants   Plants growing and propagating by self-replication of genetically identical units that can survive and function on their own if separated.

Community   Collection of populations living in the same habitat.

Crustose   Lichens which grow very close to the substrate and can only be removed by scraping.

Detritivores   An animal which feeds on dead material (often plant remains such as leaf litter, but may also be carrion).

Diapause   Period of suspended activity and development (especially in insects) often to avoid unfavourable conditions.

Dioecious   Male and female flowers occur on separate plants.

Ecotone   Well-defined transition zone between different communities.

Ecotype   A population of a widespread species which is adapted to local conditions (such as heavy metal pollution).

Eutrophic   Nutrient-rich, often used in the context of open waters.

Foliose   Lichens which are leaf-like and can be removed from the substrate with a knife.

Fruticose   Lichens which are shrubby and are either not attached to the substrate, or only attached at one point.

Heterotrophic   Utilising other organisms (living or dead) as a source of carbon.

Hydrosere   A series of communities which reflect developmental changes with increasing distance (either in space or time) from soils submerged in freshwater.

Leguminous   Belonging to the pea family, once called the Leguminosae (now the Fabaceae); a term often used to cover the plant's relationship with nitrogen-fixing bacteria.

Locally distributed   Not widely distributed.

Morph   A form of a species (often relating to colour).

Oligotrophic   Nutrient-poor, often used in the context of open waters.

Parasitoid   Insect (usually Hymenoptera or Diptera) with free-living adults that lay their eggs in another invertebrate (usually an insect larva, or sometimes a young spider) in which they develop, emerging as adults after eventually killing the host.

Parthenogenetic   Reproduction from females without male fertilisation (especially in invertebrates).

Population   Group of organisms of the same species occupying the same habitat.

Reclamation   Winning back land from a wild or waste condition to a use which was not its original one (see also restoration).

Restoration   Converting land in a wild or waste condition to its original use (see also reclamation).

Rhizome   A root-like stem on the surface or under the ground which is a means of propagation.

Rosette growth   Plant growth form, where leaves spread around the plant at ground level.

Ruderal   Tolerant of high disturbance, especially in the presence of nutrients and absence of competition.

SBI   (Site of Biological Importance) together with SINC and SNCI are local designations for sites which are protected (but usually not fully: see SSSI) because of their importance for nature conservation.

Seed bank   Accumulation of viable seeds (usually in the soil) which provide potential sources of plants in the future.

Semi-improved grassland   Grassland with some agricultural development (see also unimproved grassland).

SINC   (Site of Importance for Nature Conservation) see SBI.

SNCI   (Site of Nature Conservation Importance) see SBI.

SSSI   (Site of Special Scientific Interest) an area designated as being of particular interest because of its natural history or geology and subject to special regulations for management or development.

Succession   Sequential change in organisms and communities in response to environmental changes caused either by external factors (e.g. siltation of ponds) or the organisms themselves.

Symbiotic   Individuals of different species living together in close association.

Synanthropic   Living in close association with humans.

Taxon   (Plural: taxa) any group used in classification.

Transpiration   Evaporation of water from the pores of leaves and stems.

Trophic level   A point along a food chain (e.g. from plants to herbivores to carnivores).

Umbel   Flowers in a flat or rounded cluster.

Unimproved grassland   Grassland not used for agriculture or other developments, and therefore without applications of agrochemicals such as fertilisers (see also semi-improved grassland).

SPECIES LIST

•

This section lists the species mentioned in the book, drawing them together into their appropriate taxonomic groupings. Organisms (except viruses) are first divided into one of five kingdoms (Table 6.1). There are several more subdivisions: the major levels used here are in Table 6.2. The

Table 6.1 Summary of the five kingdoms

Kingdom	Organisms	Characteristics
Prokaryotae (Monera)	Bacteria and Cyanobacteria	Single-celled, prokaryotic (lack a membrane-bounded nucleus)
Protoctista	Nucleated algae (including seaweeds), protozoa and slime moulds	Those eukaryotes (i.e. possess a membrane-bounded nucleus) which are not fungi, plants or animals. Often single-celled, mainly aquatic (including in damp environments and the tissues of other species), often autotrophic
Fungi	Fungi and lichens	Eukaryotic, mainly multicellular, develop directly from spores with no embryological development, heterotrophic and often saprotrophic (feed on non-living organic matter)
Plantae	Plants including mosses, ferns, liverworts, conifers and flowering plants	Eukaryotic, multicellular, develop from an embryo (multicellular young organisms supported by maternal tissue), often photoautotrophic (autotrophs obtaining energy from sunlight)
Animalia	Invertebrate and vertebrate animals	Eukaryotic, multicellular, develop from a blastula (hollow ball of cells), heterotrophic

Source: Margulis and Schwartz (1988)

Table 6.2 Taxonomic levels used in species list

KINGDOM
PHYLUM
CLASS
   Order
      Family
         Species Authority [Common Name]

final division is into species. The naming of organisms in a standard way (following international convention) enables everyone to be certain which species is under consideration.

Every species is identified by a unique scientific name consisting of a binomial term (two words, the first is the genus and the second is the species). Sometimes, especially where there are several species which are difficult to separate, a different term (e.g. agg. or sect.) is placed after the genus, to indicate that several species are involved. In other cases, the name may include an x indicating that the organism (usually a plant) is a cross between two different species. A third term may be used to designate a sub-species or a variety of the species. Note that species and sub-species names are Latinised and should be shown in italics, with the generic name beginning with a capital letter. The species name is followed by the name of the person who originally described it (the authority). Where the authority is well known (e.g. if they have named many species), an abbreviation may be used (e.g. L. for Linnaeus, F. for Fabricius). An authority in parentheses indicates that the original species name has been altered (e.g. if the organism has been placed in a different genus). In the absence of parentheses around the authority, the original name is still in use today.

The names used here follow a variety of sources depending on the taxonomic group including: Hill, Preston and Smith (1991–94) for mosses and liverworts; Stace (1997) for plants; Abbott and Boss (1989) for molluscs; Oliver and Meecham (1993) for woodlice; Roberts (1985–1987) for spiders; Kloet and Hinks (1964–1975) for insects; Maitland and Lyle (1991) for fish; Cramp (1977–94) for birds; Corbet and Harris (1991) for mammals, and Hayward and Ryland (1990) for marine organisms.

PROKARYOTAE
CYANOBACTERIA (Blue-green Algae)
CYANOPHYCEAE
  Nostocales
    Oscillatoriaceae
      Oscillatoria Vaucher species
    Nostocaceae
      Nostoc Vaucher species
NITROGEN-FIXING AEROBIC BACTERIA
EUBACTERIA
  Gram-negative aerobic rods and cocci
    Rhizobiaceae
      Rhizobium Frank species
PROTOCTISTA
CHLOROPHYTA (Green Algae)
CHLOROPHYCEAE
  Chaetophorales
    Pleurococcaceae
      Desmococcus olivaceus (Pers. ex Ach.)
  Ulvales
    Prasiolaceae
      Prasiola crispa (Lightf.)
    Clorococcaceae
      Trebouxia Puymaly species
FUNGI
BASIDIOMYCOTA (Mushrooms and Toadstools)

HYMENOMYCETES
  Agaricales
    Pleurotaceae
      Lentinus lepideus (Fr.)
    Pluteaceae
      Volvariella speciosa Singer
    Cortinariaceae
      Tubaria furfuracea (Pers.)
    Coprinaceae
      Coprinus bisporus Lange
      Coprinus cinerus (Sch.)
      Coprinus comatus (Müller) [Shaggy Ink Cap]
  Exobasidiales
    Dacrymycetaceae
      Dacrymyces deliquescens (Nees)
DISCOMYCETES
  Rhytismatales
    Rhytismataceae
      Rhytisma acerinum (Pers.) [Tar-spot Fungus]
MYCOPHYCOPHYTA (Lichens)
ASCOMYCETES
  Graphidales
    Graphidaceae
      Graphis elegans (Borrer. ex Sm.)
  Gyalectales
    Gyalectaceae

Dimerella lutea (Dicks.)
Pachyphiale cornea (With.)

Lecanorales

Agyriaceae
Trapelia Choisy species

Biatoraceae
Lecania erysibe (Ach.)

Candelariaceae
Candelariella aurella (Hoffm.)

Lecanoraceae
Lecanora chlarotera Nyl.
Lecanora conizaeoides Nyl.
Lecanora dispersa (Pers.)
Scoliciosporum umbrinum (Ach.)

Parmeliaceae
Foraminella ambigua (Wulf.) (Wulf.)
Hypogymnia physodes (L.)
Parmelia caperata (L.)
Parmelia glabratula (Lamy)
Parmelia perlata (Huds.)
Parmelia reticulata Tayl.
Parmelia saxatilis (L.)
Parmelia subrudecta Nyl.
Parmelia sulcata Taylor
Usnea articulata (L.)
Usnea ceratina Ach.
Usnea filipendula Stirton [Beard-moss]
Usnea florida (L.)
Usnea subfloridana Stirton

Peltigaraceae
Peltigera didactyla (Nyl. ex Vain)

Lobariaceae
Lobaria amplissima (Scop.)
Lobaria pulmonaria (L.)
Lobaria scrobiculata (Scop.)
Sticta limbata (Sm.)

Pannariaceae
Pannaria Del. species

Pertusariaceae
Pertusaria hemisphaerica (Flörke)

Teloschistaceae
Caloplaca citrina (Hoffm.)
Teloschistes flavicans (Swartz)
Xanthoria (Fr.) species

Imperfect Lichenized Fungi
Lepraria incana (L.)

PLANTAE (plants)
BRYOPHYTA (Mosses and Liverworts)
HEPATICAE (Liverworts)
Marchantiales

Lunulariaceaea
Lunularia crutiata (L.)

Marchantiaceae
Marchantia polymorpha L.

MUSCI (Mosses)
Sphagnales

Sphagnaceae
Sphagnum L. species

Dicranales

Dicranaceae
Ceratodon purpureus (Hedw.) [Purple-fruiting Heath-moss]

Encalyptales

Encalyptaceae
Encalypta streptocarpa Hedw. [Extinguisher-moss]

Pottiales

Pottiaceae
Tortula muralis Hedw. [Wall-screw Moss]
Barbula Hedw. species [Beard Mosses]

Funariales

Funariaceae
Funaria hygrometrica Hedw. [Common Cord Moss]

Bryales

Bryaceae
Leptobryum pyriforme (Hedw.)
Bryum argenteum (Hedw.) [Silvery Thread-moss]
Bryum capillare Hedw.

SPHENOPHYTA
EQUISETOPSIDA (Horsetails)
Equisetaceae (Horsetail family)
Equisetum arvense L. [Field Horsetail]

FILICINOPHYTA (Ferns)
PTEROPSIDA
Polypodiaceae (Polypody family)
Polypodium vulgare L. [Polypody]

Dennstaedtiaceae (Bracken family)
Pteridium aquilinum (L.) [Bracken]

Aspleniaceae (Spleenwort family)
Asplenium adiantum-nigrum L. [Black Spleenwort]
Asplenium ruta-muraria L. [Wall-rue]
Asplenium trichomanes L. [Maidenhair Spleenwort]
Ceterach officinarum Willd. [Rustyback]

Woodsiaceae (Lady-fern family)
Athyrium felix-femina (L.) [Lady-fern]

Dryopteridaceae (Buckler-fern family)

Dryopteris dilatata (Hoffm.) [Broad Buckler-fern]

Dryopteris felix-mas (L.) [Male-fern]

Azollaceae (Water Fern family)

Azolla filiculoides Lam. [Water Fern]

CONIFEROPHYTA (Conifers)

PINOPSIDA

Pinaceae (Pine family)

Picea abies (L.) [Norway Spruce]

Pinus sylvestris L. [Scots Pine]

Cupressaceae (Juniper family)

Juniperus communis L. [Common Juniper]

Juniperus virginiana L. [Pencil Cedar]

Araucariaceae (Monkey-puzzle family)

Araucaria araucana (Molina) [Monkey-puzzle]

Taxaceae (Yew family)

Taxus baccata L. [Yew]

ANGIOSPERMOPHYTA (Flowering Plants)

MAGNOLIOPSIDA (Dicotyledons)

Nymphaeales

Ceratophyllaceae (Hornwort family)

Ceratophyllum submersum L. [Soft Hornwort]

Ranunculales

Ranunculacea (Buttercup family)

Anenome nemorosa L. [Wood Anenome]

Ranunculus ficaria L. [Lesser Celandine]

Ranunculus lingua L. [Greater Spearwort]

Ranunculus repens L. [Creeping Buttercup]

Berberidaceae (Barberry family)

Mahonia Nutt. species [Oregon-grapes]

Berberis vulgaris L. [Barberry]

Papaverales

Fumariaceae (Fumitory family)

Pseudofumaria lutea (L.) [Yellow Corydalis]

Hamamelidales

Platanaceae (Plane family)

Platanus x hispanica Mill. ex Muenchh. [London plane]

Urticales

Ulmaceae (Elm family)

Ulmus glabra Huds. [Wych Elm]

Moraceae (Mulberry family)

Ficus carica L. [Fig]

Urticaceae (Nettle family)

Urtica dioica L. [Common Nettle]

Soleirolia soleirolii (Req.) [Mind-your-own-business]

Myricales

Myricaceae (Bog-myrtle family)

Myrica gale L. [Bog-myrtle]

Fagales

Fagaceae (Beech family)

Fagus sylvatica L. [Beech]

Castanea sativa Mill. [Sweet Chestnut]

Quercus petraea (Matt.) [Sessile Oak]

Quercus robur L. [Pedunculate Oak]

Quercus rubra L. [Red Oak]

Betulaceae (Birch family)

Betula pendula Roth [Silver Birch]

Betula pubescens Ehrh. [Downy Birch]

Alnus glutinosa (L.) [Alder]

Alnus incana (L.) [Grey Alder]

Carpinus betulus L. [Hornbeam]

Corylus avellana L. [Hazel]

Caryophyllales

Chenopodiaceae (Goosefoot family)

Chenopodium album L. [Fat-hen]

Chenopodium rubrum L. [Red Goosefoot]

Atriplex patula L. [Common Orache]

Atriplex portulacoides L. [Sea-purslane]

Atriplex prostrata Boucher ex DC. [Spear-leaved Orache]

Salicornia L. species [Glassworts]

Caryophyllaceae (Pink family)

Stellaria media (L.) [Common Chickweed]

Cerastium fontanum Baumg. [Common Mouse-ear]

Sagina procumbens L. [Procumbent Pearlwort]

Spergula arvensis L. [Corn Spurrey]

Spergularia marina (L.) [Lesser Sea-spurrey]

Silene vulgaris Garcke [Bladder Campion]

Dianthus barbatus L. [Sweet-William]

Polygonales

Polygonaceae (Knotweed family)

Polygonum aviculare L. [Knotgrass]

Fallopia japonica (Houtt.) [Japanese Knotweed]

Rumex acetosella L. [Sheep's Sorrel]

Rumex conglomeratus Murray [Clustered Dock]

Rumex crispus L. [Curled Dock]

Rumex pulcher L. [Fiddle Dock]

Plumbaginales

Plumbaginaceae (Thrift family)

Limonium vulgare Mill. [Common Sea-lavender]

Armeria maritima Willd. [Thrift]

Theales

Clusiaceae (St John's-wort family)

Hypericum perforatum L. [Perforate St John's-wort]

Malvales
  Tiliaceae (Lime family)
    Tilia cordata Mill. [Small-leaved Lime]
    Tilia x europaea L. [Lime]
  Malvaceae (Mallow family)
    Malva sylvestris L. [Common Mallow]
Violales
  Violaceae (Violet family)
    Viola lutea Huds. [Mountain Pansy]
    Viola odorata L. [Sweet Violet]
    Viola riviniana Rchb. [Common Dog-violet]
Salicales
  Salicaceae (Willow family)
    Populus nigra L. [Black-poplar]
    Populus tremula L. [Aspen]
    Salix alba L. [White Willow]
    Salix alba vitellina (L.) [Golden Willow]
    Salix aurita L. [Eared Willow]
    Salix caprea L. [Goat Willow]
    Salix cinerea L. [Grey Willow]
    Salix daphnoides Vill. [European Violet-willow]
Capperales
  Brassicaceae (Cabbage family)
    Sisymbrium L. species [Rockets]
    Sisymbrium officianale (L.) [Hedge Mustard]
    Arabidopsis thaliana (L.) [Thale Cress]
    Erysimum cheiri (L.) [Wallflower]
    Rorippa nasturtium-aquaticum (L.) [Water-cress]
    Rorippa sylvestris (L.) [Creeping Yellow-cress]
    Erophila verna (L.) [Common Whitlowgrass]
    Capsella bursa-pastoris (L.) [Shepherd's-purse]
    Diplotaxis DC. species [Wall-rockets]
    Brassica napus L. [Oil-seed Rape]
    Sinapis arvensis L. [Charlock]
    Crambe maritima L. [Sea-kale]
  Resedaceae (Mignonette family)
    Reseda luteola L. [Weld]
Ericales
  Empetraceae (Crowberry family)
    Empetrum nigrum L. [Crowberry]
  Ericaceae (Heather family)
    Rhododendron ponticum L. [Rhododendron]
    Calluna vulgaris (L.) [Heather]
    Erica cinerea L. [Bell Heather]
    Erica tetralix L. [Cross-leaved Heath]
    Vaccinium myrtillus L. [Bilberry]
Primulales
  Primulaceae (Primrose family)

Primula veris L. [Cowslip]
Primula vulgaris Huds. [Primrose]
Lysimachia thyrisiflora L. [Tufted Loosestrife]
Anagallis arvensis arvensis L. [Scarlet Pimpernel]

Rosales
  Grossulariaceae (Gooseberry family)
    Ribes sanguineum Pursh [Flowering Currant]
    Ribes uva-crispa L. [Gooseberry]
  Crassulaceae (Stonecrop family)
    Sedum L. species [Stonecrops]
  Saxifragaceae (Saxifrage family)
    Bergenia crassifolia (L.) [Elephant-ears]
    Saxifraga granulata L. [Meadow Saxifrage]
  Rosaceae (Rose family)
    Filipendula ulmaria (L.) [Meadowsweet]
    Rubus chamaemorus L. [Cloudberry]
    Rubus sect. *Glandulosus* Wimm. and Grab. [Brambles]
    Fragaria vesca L. [Wild Strawberry]
    Fragaria x ananassa (Duchesne) [Garden Strawberry]
    Geum urbanum L. [Wood Avens]
    Prunus avium (L.) [Wild Cherry]
    Prunus laurocerasus L. [Cherry Laurel]
    Prunus lusitanica L. [Portugal Laurel]
    Prunus spinosa L. [Blackthorn]
    Pyrus communis L. [Pear]
    Malus domestica Bork. [Apple]
    Sorbus aria (L.) [Common Whitebeam]
    Sorbus aucuparia L. [Rowan]
    Sorbus intermedia (Ehrh.) [Swedish Whitebeam]
    Cotoneaster Medik. species [Cotoneasters]
    Crataegus monogyna Jacq. [Hawthorn]
    Crataegus laevigata (Poir.) [Midland Hawthorn]
Fabales
  Fabaceae (Pea family)
    Galega officinalis L. [Goat's-rue]
    Lotus corniculatus L. [Common Bird's-foot-trefoil]
    Vicia cracca L. [Tufted Vetch]
    Lathyrus latifolius L. [Broad-leaved Everlasting-pea]
    Melilotus altissimus Thuill. [Tall Melilot]
    Medicago lupulina L. [Black Medick]
    Trifolium pratense L. [Red Clover]
    Trifolium repens L. [White Clover]
    Lupinus polyphyllus Lindl. [Garden Lupin]
    Laburnum anagyroides Medik. [Laburnum]
    Cytisus scoparius (L.) [Broom]

Ulex europaeus L. [Gorse]
Haloragales
Haloragaceae (Water-milfoil family)
Myriophyllum spicatum L. [Spiked Water-milfoil]
Myrtales
Onagraceae (Willowherb family)
Epilobium ciliatum Raf. [American Willowherb]
Epilobium hirsutum L. [Great Willowherb]
Chamerion angustifolium (L.) [Rosebay Willowherb]
Celastrales
Aquifoliaceae (Holly family)
Ilex aquifolium L. [Holly]
Euphorbiales
Euphorbiaceae (Spurge family)
Mercurialis perennis L. [Dog's Mercury]
Linales
Linaceae (Flax family)
Linum catharticum L. [Fairy Flax]
Sapindales
Hippocastanaceae (Horse-chestnut family)
Aesculus hippocastanum L. [Horse-chestnut]
Aceraceae (Maple family)
Acer campestre L. [Field Maple]
Acer palmatum Thunb. [Japanese Maple]
Acer platanoides L. [Norway Maple]
Acer pseudoplatanus L. [Sycamore]
Rutaceae (Rue family)
Citrus L. species [Citrus]
Geraniales
Oxalidaceae (Wood-sorrel family)
Oxalis acetosella L. [Wood-sorrel]
Geraniaceae (Crane's-bill family)
Geranium robertianum L. [Herb-robert]
Balsaminaceae (Balsam family)
Impatiens L. species [Balsams]
Impatiens glandulifera Royle [Indian Balsam]
Apiales
Araliaceae (Ivy family)
Hedera helix L. [Ivy]
Apiaceae (Carrot family)
Coriandrum sativum L. [Coriander]
Conopodium majus (Gouan) [Pignut]
Pimpinella saxifraga L. [Burnet-saxifrage]
Aegopodium podagraria L. [Ground-elder]
Berula erecta (Huds.) [Lesser Water-parsnip]
Foeniculum vulgare Mill. [Fennel]
Cuminum cyminum L. [Cumin]
Pastinaca sativa sativa (Mill.) [Wild Parsnip]
Heracleum mantegazzianum Sommier and

Levier [Giant Hogweed]
Heracleum sphondylium L. [Hogweed]
Daucus carota sativus (Hoffm.) [Carrot]
Gentianales
Gentianaceae (Gentian family)
Centaurium erythraea Rafn. [Common Centaury]
Blackstonia perfoliata (L.) [Yellow-wort]
Solanales
Solanaceae (Nightshade family)
Lycopersicon esculentum Mill. [Tomato]
Solanum nigrum L. [Black Nightshade]
Solanum tuberosum L. [Potato]
Convolvaceae (Bindweed family)
Convolvulus arvensis L. [Field Bindweed]
Calystegia sepium (L.) [Hedge Bindweed]
Calystegia silvatica (Kit.) [Large Bindweed]
Polemoniaceae (Jacob's-ladder family)
Polemonium caeruleum L. [Jacob's Ladder]
Lamiales
Lamiaceae (Deadnettle family)
Lamiastrum galeobdolon (L.) [Yellow Archangel]
Prunella vulgaris L. [Selfheal]
Calamintha calamintha (L.) [Lesser Calamint]
Lycopus europaeus L. [Gypsywort]
Mentha pulegium L. [Pennyroyal]
Plantaginales
Plantaginaceae (Plantain family)
Plantago lanceolata L. [Ribwort Plantain]
Plantago major L. [Greater Plantain]
Plantago maritima L. [Sea Plantain]
Scrophulariales
Buddlejaceae (Butterfly-bush family)
Buddleja davidii Franch. [Butterfly-bush]
Oleraceae (Ash family)
Fraxinus excelsior L. [Ash]
Lingustrum vulgare L. [Wild Privet]
Scrophulariaceae (Figwort family)
Scrophularia auriculata L. [Water Figwort]
Limosella aquatica L. [Mudwort]
Antirrhinum majus L. [Snapdragon]
Chaenorhinum minus (L.) [Small Toadflax]
Cymbalaria muralis P. Gaertn., B. Mey. and Scherb. [Ivy-leaved Toadflax]
Digitalis purpurea L. [Foxglove]
Veronica L. species [Speedwells]
Veronica beccabunga L. [Brooklime]
Euphrasia L. species [Eyebrights]

Rubiales
  Rubiaceae (Bedstraw family)
    Galium aparine L. [Cleavers]
    Galium verum L. [Lady's Bedstraw]
Dipsacales
  Caprifoliaceae (Honeysuckle family)
    Sambucus nigra L. [Elder]
  Valerianaceae (Valerian family)
    Centranthus ruber (L.) [Red Valerian]
Asterales
  Asteraceae (Daisy family)
    Cirsium arvense (L.) [Creeping Thistle]
    Cirsium vulgare (Savi) [Spear Thistle]
    Centaurea nigra L. [Common Knapweed]
    Hypochaeris radicata L. [Cat's-ear]
    Leontodon hispidus L. [Rough Hawkbit]
    Tragopogon pratensis L. [Goat's-beard]
    Sonchus asper (L.) [Prickly Sow-thistle]
    Sonchus oleraceus L. [Smooth Sow-thistle]
    Lactuca saligna L. [Least Lettuce]
    Taraxacum Wigg. sect. [Dandelions]
    Crepis capillaris (L.) [Smooth Hawk's-beard]
    Pilosella officinarum F. W. Schulz and Sch. Bip. [Mouse-ear-hawkweed]
    Hieracium L. sect. [Hawkweeds]
    Inula conyzae (Griess.) [Ploughman's-spikenard]
    Pulicaria vulgaris Gaertn. [Small Fleabane]
    Solidago virgaurea L. [Goldenrod]
    Aster novi-belgii L. [Confused Michaelmas-daisy]
    Aster tripolium L. [Sea Aster]
    Bellis perennis L. [Daisy]
    Tanacetum parthenium (L.) [Feverfew]
    Tanacetum vulgare L. [Tansy]
    Artemisia absinthium L. [Wormwood]
    Artemisia vulgaris L. [Mugwort]
    Achillea millefolium L. [Yarrow]
    Leucanthemum vulgare Lam. [Oxeye Daisy]
    Leucanthemum x superbum (Bergmans ex J. W. Ingram) [Shasta Daisy]
    Matricaria discoidea DC. [Pineappleweed]
    Tripleurospermum inordorum (L.) [Scentless Mayweed]
    Senecio jacobaea L. [Common Ragwort]
    Senecio paludosus L. [Fen Ragwort]
    Senecio squalidus L. [Oxford Ragwort]
    Senecio viscosus L. [Sticky Groundsel]
    Senecio vulgaris L. [Groundsel]
    Tussilago farfara L. [Colt's-foot]
    Bidens L. species [Bur-marigolds]
LILIIDAE (Monocotyledons)
  Alismatales
    Alismataceae (Water-plantain family)
      Luronium natans (L.) [Floating water-plantain]
      Alisma plantago-aquatica L. [Water-plantain]
  Hydrocharitales
    Hydrocharitaceae (Frogbit family)
      Elodea canadensis Michx. [Canadian Waterweed]
  Najadales
    Potamogetonaceae (Pondweed family)
      Potamogeton berchtoldii Fieber [Small Pondweed]
      Potamogeton gramineus L. [Various-leaved Pondweed]
      Potamogeton natans L. [Broad-leaved Pondweed]
  Arales
    Lemnaceae (Duckweed family)
      Lemna gibba L. [Fat Duckweed]
      Lemna minor L. [Common Duckweed]
  Juncales
    Juncaceae (Rush family)
      Juncus L. species [Rushes]
  Cyperales
    Cyperaceae (Sedge family)
      Eriophorum angustifolium Honck. [Common Cottongrass]
      Bolboschoenus maritimus (L.) [Sea Club-rush]
      Carex divisa Huds. [Divided Sedge]
      Carex flacca Schreb. [Glaucous Sedge]
    Poaceae (Grass family)
      Festuca ovina L. [Sheep's-fescue]
      Festuca rubra L. agg. [Red Fescue]
      Lolium perenne L. [Perennial Rye-grass]
      Cynosurus cristatus L. [Crested Dog's-tail]
      Puccinellia distans (Jacq.) [Reflexed Saltmarsh-grass]
      Puccinellia maritima (Huds.) [Common Saltmarsh-grass]
      Briza media L. [Quaking-grass]
      Poa annua L. [Annual Meadow-grass]
      Poa pratensis L. [Smooth Meadow-grass]
      Poa trivialis L. [Rough Meadow-grass]
      Dactylis glomerata L. [Cock's-foot]
      Catapodium rigidum (L.) [Fern-grass]
      Glyceria maxima (Hartm.) [Reed Sweet-grass]
      Arrhenatherum elatius (L.) [False Oat-grass]
      Trisetum flavescens (L.) [Yellow Oat-grass]

Holcus lanatus L. [Yorkshire-fog]
Phalaris arundinacea L. [Reed canary-grass]
Agrostis capillaris L. [Common Bent]
Agrostis stolonifera L. [Creeping Bent]
Ammophila arenaria (L.) [Marram]
Anisantha sterilis (L.) [Barren Brome]
Brachypodium sylvaticum (Huds.) [False Brome]
Elytrigia repens (L.) [Common Couch]
Hordeum murinum L. [Wall Barley]
Phragmites australis (Cav.) [Common Reed]
Spartina alterniflora Loisel. [Smooth Cord-grass]
Spartina anglica C. E. Hubb. [Common Cord-grass]
Spartina maritima (Curtis) [Small Cord-grass]

Typhales
Sparganiaceae (Bur-reed family)
Sparganium angustifolium Michx. [Floating Bur-reed]
Sparganium erectum L. [Branched Bur-reed]
Typhaceae (Bulrush family)
Typha latifolia L. [Bulrush]

Liliales
Liliaceae (Lily family)
Hyacinthoides non-scripta (L.) [Bluebell]
Allium ursinum L. [Ransoms]
Iridaceae (Iris family)
Iris pseudacorus L. [Yellow Iris]

Orchidales
Orchidaceae (Orchid family)
Dactylorhiza incarnata (L.) [Early Marsh-orchid]
Dactylorhiza praetermissa (Druce) [Southern Marsh-orchid]
Dactylorhiza purpurella (T. and T. A. Stephenson) [Northern Marsh-orchid]

ANIMALIA (animals)
ROTIFERA (Rotifers)
NEMATODA (Nematode worms or Roundworms)
ECTOPROCTA (Bryozoa)
Conopeum seurati (Canu)
MOLLUSCA
GASTROPODA
Archaeogastopoda
Hydrobiidae
Hydrobia ulvae (Pennant) [Laver Spire Shell]
Basommatophora
Lymnaeidae
Lymnaea stagnalis (L.) [Great Pond Snail]
Stylommatophora
Pyramidulidae

Pyramidula rupestris (Draparnaud) [Rock Snail]
Clausiliidae
Balea perversa (Linné) [Tree Snail]
Limacidae
Limax flavus Linné [Yellow Slug]
Deroceras reticulatum (Müller)
Helicidae
Arianta arbustorum (Linné)
Cepeae nemoralis (Linné)
Cepeae hortensis (Müller)
BIVALVIA (Bivalves)
Veneroida
Dreissenidae
Dreissena polymorpha (Pallas) [Zebra Mussel]
Pisidiidae
Pisidium L. species [Pea Mussel]
ANNELIDA (Segmented Worms)
POLYCHAETA
Arenicolidae
Arenicola marina L. [Lugworm]
OLIGOCHAETA
Haplotaxida
Lumbricidae (Earthworms)
HIRUDINEA (Leeches)
Rhynchobdellida
Glossiphoniidae
Erpobdella octoculata L.
TARDIGRADA (Tardigrades or Water Bears)
ARTHROPODA (Arthropods)
MALACOSTRACA
Amphipoda (Amphipods)
Gammaridae
Gamerus pulex (L.) [Freshwater Shrimp]
Corophiidae
Corophium curvispinum Sars.
Corophium insidiosum Crawford
Corophium volutator (Pallas)
Isopoda (Woodlice)
Asellidae
Asellus aquaticus L. [Water Hoglouse]
Trichoniscidae
Androniscus dentiger Verhoeff [Rosy Woodlouse]
Trichoniscus pusillus Brandt [Common Pygmy Woodlouse]
Oniscidae
Oniscus asellus L. [Common Shiny Woodlouse]
Philosciidae
Philoscia muscorum (Scopoli) [Common Striped Woodlouse]

Armadillidiidae
Armadillidium album Dollfus
Armadillidium vulgare (Latreille) [Common Pill Woodlouse]
Porcellionidae
Porcellio scaber Latreille [Common Rough Woodlouse]
Porcellio spinicornis Say
Decapoda
Palaemonidae
Palaemonetes varians (Leach)
ARACHNIDA
Acarina (Mites and Ticks)
Argasidae
Argas reflexus (Fabricius) [Pigeon Tick]
Scorpionida (Scorpions)
Chactidae
Euscorpius flavicaudis (de Geer)
Opiliones (Harvestmen)
Araneae (Spiders)
Amaurobiidae
Amaurobius ferox (Walckenaer)
Dysderidae
Dysdera crocota Koch [Woodlouse Spider]
Segestriidae
Segestria senoculata (L.)
Pholcidae
Pholcus phalangioides (Fuesslin)
Salticidae
Salticus scenicus (Clerck) [Zebra Spider]
Lycosidae (Wolf Spiders)
Agelinidae
Tegenaria duellica Simon
Araneidae
Araneus diadematus Clerck [Garden Spider]
Cyclosa conica (Pallas) [Orb-web-spinning Spider]
DIPLOPODA (Millipedes)
CHILOPODA (Centipedes)
INSECTA (Insects)
Collembola (Springtails)
Odonata (Dragonflies and Damselflies)
Coenagrionidae
Ischnura elegans (Van der Linden) [Blue-tailed Damselfly]
Lestidae
Lestes dryas Kirby [Scarce Emerald Dragonfly]
Aeshnidae
Aeshna mixta Latreille [Migrant Hawker]
Libellulidae

Sympetrum sanguineum (Müller) [Ruddy Darter]
Sympetrum striolatum (Charpentier) [Common Darter]
Plecoptera (Stoneflies)
Ephemoptera (Mayflies)
Orthoptera (Grasshoppers and Crickets)
Gryllidae (Crickets)
Acheta domesticus (L.) [House Cricket]
Acrididae (Grasshoppers)
Chorthippus brunneus (Thunberg) [Field Grasshopper]
Chorthippus parallelus (Zetterstedt) [Meadow Grasshopper]
Dictyoptera (Cockroaches)
Blattidae
Periplaneta americana (L.) [American Cockroach]
Blatta orientalis L. [Common Cockroach]
Blatellidae
Blattella germanica (L.) [German Cockroach]
Hemiptera (True Bugs)
Anthocoridae
Anthocoris nemorum (L.)
Gerridae (Pond Skaters)
Notonectidae
Notonecta L. species [Backswimmers]
Corixiae (Water Boatmen)
Aphididae
Impatientinum balsamines (Kaltenbach)
Aphis fabae (Scopoli) [Black Bean Aphid]
Drepanosiphum platanoidis (Shrank) [Sycamore Aphid]
Delphacidae
Asiraca clavicornis (F.)
Trichoptera (Caddis flies)
Neuroptera
Sialidae (Alder Flies)
Hemerobiidae (Brown Lacewings)
Chrysopidae (Green Lacewings)
Diptera (Two-winged or True Flies)
Tipulidae (Crane Flies)
Chironomidae (Non-biting Midges)
Syrphidae (Hoverflies)
Episyrphus balteatus (Degeer)
Eristalis tenax L. [Dronefly or Rat-tailed Maggot]
Sphaerophoria scripta (L.)
Syrphus ribesii (L.)
Coleoptera (Beetles)
Carabidae (Ground Beetles)

Nebria brevicollis (F.)
Nebria complanata (L.)
Broscus cephalotes (L.)
Dytiscidae (Water Beetles)
Staphylinidae (Rove Beetles)
Staphylinus olens (Müller) [Devil's Coach-horse Beetle]
Coccinellidae (Ladybirds)
Adalia bipunctata (L.) [Two-spot Ladybird]
Coccinella septempunctata L. [Seven-spot Ladybird]
Chrysomelidae (Leaf Beetles)
Curculionidae (Weevils)
Hymenoptera (Bees, Wasps, Ants and Allies)
Formicidae (Ants)
Ponera coarctata (Latreille)
Lasius flavus (F.) [Yellow Meadow Ant]
Lasius niger (L.) [Black Garden Ant]
Vespidae (Wasps)
Vespa germanica (F.) [German Wasp]
Apidae (Bees)
Halictus Latreille species [Mining Bees]
Andrena F. species [Mining Bees]
Megachile willughbiella (Kirby) [Rose-leaf Cutter Bees]
Osmia rufa (L.) [Mason Bee]
Bombus pratorum (L.) [Bumble Bee]
Bombus pascuorum (Scopoli) [Bumble bee]
Apis mellifera L. [Honey Bee]
Lepidoptera (Butterflies and Moths)
Nymphalidae
Inachis io L. [Peacock]
Vanessa atalanta L. [Red Admiral]
Aglais urticae L. [Small Tortoiseshell]
Satyridae (Brown Butterflies)
Maniola jurtina L. [Meadow Brown]
Lasiommata megera L. [Wall Brown]
Lycaenidae (Blue Butterflies and Allies)
Lycaena phlaeas L. [Small Copper]
Celastrina argiolus L. [Holly Blue]
Polyommatus icarus Rottemburg [Common Blue]
Hesperiidae (Skippers)
Thymelicus flavus Poda [Small Skipper]
Ochiodes venatus Turati [Large Skipper]
Zygaenidae
Zygaena filipendulae (L.) [Six-spot Burnet]
Geometridae
Biston betularia (L.) [Peppered Moth]
Sphingidae (Hawkmoths)

Deilephila elpenor (L.) [Elephant Hawk-moth]
Arctiidae
Tyria jacobaeae L. [The Cinnabar]
Noctuidae
Orthosia gothica L. [Hebrew Character]
Agrochola circellaris Hufn. [The Brick]
Diachrysia chrysitis L. [Burnished Brass]
Autographa gamma L. [Silver Y]
Hypena proboscidalis L. [The Snout]
CHORDATA
OSTEICHTHYES (Bony Fish)
Isospondyli
Osmeridae (Smelt family)
Osmerus eperlanus (L.) [Smelt]
Haplomi
Esocidae (Pike family)
Esox lusius L. [Pike]
Ostariophysi
Cyprinidae (Carp family)
Cyprinus carpio L. [Common Carp]
Carassius carassius L. [Crucian Carp]
Tinca tinca (L.) [Tench]
Blicca bjoerkna (L.) [Silver Bream]
Abramis brama (L.) [Common Bream]
Alburnus alburnus (L.) [Bleak]
Scardinius erythrophthalmus (L.) [Rudd]
Rutilus rutilus (L.) [Roach]
Leuciscus cephalus (L.) [Chub]
Leuciscus idus (L.) [Orfe]
Apodes
Anguillidae (Eel family)
Anguilla anguilla (L.) [European Eel]
Thoracostei
Gasterosteidae (Stickleback family)
Gasterosteus aculeatua L. [Three-spined Stickleback]
Pungitius pungitius (L.) [Nine-spined Stickleback]
Anacanthini
Gadidae (Cod family)
Lota lota (L.) [Burbot]
Perciformes
Percidae (Perch family)
Perca fluviatilis L. [Perch]
Gymnocephalus cernua (L.) [Ruffe]
Stizostedion lucioperca (L.) [Pikeperch or Zander]
Scleroparei
Cottidae (Sculpin family)
Cottus gobio L. [Bullhead]

AMPHIBIA (Amphibians)
  Caudata
    Salamandridae (Newts and Salamanders)
      Triturus cristatus (Laurenti) [Great-crested Newt]
      Triturus helveticus (Razoumowski) [Palmate Newt]
      Triturus vulgaris (L.) [Smooth Newt]
      Triturus alpestris Laurenti [Alpine Newt]
  Salientia
    Bufonidae (Toads)
      Bufo bufo (L.) [Common Toad]
      Bufo calamita Laurenti [Natterjack Toad]
    Ranidae (Frogs)
      Rana temporaria L. [Common Frog]
REPTILIA (Reptiles)
  Squamata
    Anguidae
      Anguis fragilis L. [Slow-worm]
    Lacertidae
      Lacerta agilis L. [Sand Lizard]
      Lacerta vivipara Jacquin [Common Lizard]
AVES (Birds)
  Podicipediformes
    Podicipedidae (Grebes)
      Podiceps cristatus (L.) [Great-crested Grebe]
      Tachybaptus ruficollis (Pallas) [Little Grebe]
  Ciconiiformes
    Ardeidae (Bitterns and Herons)
      Botaurus stellaris (L.) [Bittern]
  Anseriformes
    Anatidae (Wildfowl)
      Cygnus olor (Gmelin) [Mute Swan]
      Branta canadensis (L.) [Canada Geese]
      Tadorna tadorna (L.) [Shelduck]
      Anas clypeata L. [Shoveler]
      Anas crecca L. [Teal]
      Anas penelope L. [Wigeon]
      Anas platyrhynchos L. [Mallard]
      Anas querquedula L. [Garganey]
      Anas strepera L. [Gadwall]
      Aythya ferina (L.) [Pochard]
      Aythya fuligula (L.) [Tufted Duck]
      Aythya marila (L.) [Scaup]
      Melanitta nigra (L.) [Common Scoter]
      Oxyura jamaicensis (Gmelin) [Ruddy Duck]
  Accipitriformes
    Accipitridae (Hawks and Allies)
      Accipiter nisus (L.) [Sparrowhawk]
  Falconiformes

Falconidae (Falcons and Allies)
    Falco tinnunculus L. [Kestrel]
    Falco columbarius L. [Merlin]
    Falco peregrinus Tunstall [Peregrine]
Galliformes
  Phasianidae (Partridges, Pheasants and Allies)
    Alectoris rufa (L.) [Red-legged Partridge]
    Perdix perdix (L.) [Partridge]
    Phasianus colchicus L. [Pheasant]
Gruiformes
  Rallidae (Rails)
    Gallinula chloropus (L.) [Moorhen]
    Fulica atra L. [Coot]
Charadiiformes (Plovers and Allies)
  Haematopodidae (Oystercatchers)
    Haematopus ostralegus L. [Oystercatcher]
  Charadriidae (Plovers and Lapwings)
    Charadrius dubius Scopoli [Little Ringed Plover]
    Charadrius hiaticula L. [Ringed Plover]
    Vanellus vanellus (L.) [Lapwing]
  Scolopacidae (Sandpipers and Allies)
    Gallinago gallinago (L.) [Snipe]
    Scolopax rusticola L. [Woodcock]
    Tringa nebularia (Gunnerus) [Greenshank]
    Tringa ochropus L. [Green Sandpiper]
    Tringa totanus (L.) [Redshank]
    Actitis hypoleucos L. [Common Sandpiper]
  Laridae (Gulls)
    Larus argentatus Pontoppidan [Herring Gull]
    Larus canus L. [Common Gull]
    Larus fuscus L. [Lesser Black-backed Gull]
    Larus novaehollandiae Stephens [Silver Gull]
    Larus ridibundus L. [Black-headed Gull]
  Sternidae (Terns)
    Sterna hirundo L. [Common Tern]
Columbiformes
  Columbidae (Pigeons)
    Columba livia Gmelin [Feral Pigeon]
    Columba oenas L. [Stock Dove]
    Columba palumbus L. [Wood Pigeon]
    Streptopelia decaoto (Frivaldsky) [Collared Dove]
    Streptopelia turtur (L.) [Turtle Dove]
Cuculiformes
  Cuculidae (Cuckoos)
    Cuculus canorus L. [Cuckoo]
Stringiformes
  Tytonidae (Barn Owls and Allies)
    Tyto alba (Scopoli) [Barn Owl]

Strigidae (Brown Owls and Allies)
   Athene noctua (Scopoli) [Little Owl]
   Strix aluco L. [Tawny Owl]
   Asio flammeus (Pontoppidan) [Short-eared Owl]
   Asio otus (L.) [Long-eared Owl]
Apodiformes
  Apodidae (Swifts)
   Apus apus (L.) [Swift]
Coraciiformes
  Alcedinidae (Kingfishers)
   Alcedo atthis (L.) [Kingfisher]
Piciformes
  Picidae (Woodpeckers and Allies)
   Picus viridis (L.) [Green Woodpecker]
   Dendrocops major (L.) [Great-spotted Woodpecker]
   Dendrocops minor (L.) [Lesser-spotted Woodpecker]
Passeriformes (Perching Birds)
  Alaudidae (Larks)
   Alauda arvensis L. [Skylark]
  Hirundinidae (Swallows)
   Riparia riparia (L.) [Sand Martin]
   Hirundo rustica L. [Swallow]
   Delichon urbica (L.) [House Martin]
  Motacillidae (Pipits and Wagtails)
   Anthus pratensis (L.) [Meadow Pipit]
   Motacilla alba Gould [Pied Wagtail]
   Motacilla cinerea Tunstall [Grey Wagtail]
   Motacilla flava L. [Yellow Wagtail]
  Bombycillidae (Waxwings and Hypocoliuses)
   Bombycilla garrulus (L.) [Waxwing]
  Troglodytidae (Wrens)
   Troglodytes troglodytes (L.) [Wren]
  Prunellidae (Accentors)
   Prunella modularis (L.) [Dunnock]
  Turdidae (Chats and Thrushes)
   Erithacus rubecula (L.) [Robin]
   Turdus merula L. [Blackbird]
   Turdus philomelos Brehm [Song Thrush]
   Turdus torquatus L. [Ring Ouzel]
   Turdus viscivorus L. [Mistle Thrush]
  Sylviidae (Warblers and Allies)
   Acrocephalus schoenobaenus (L.) [Sedge Warbler]
   Acrocephalus scirpaceus (Hermann) [Reed Warbler]
   Sylvia atricapilla (L.) [Black Cap]
   Sylvia borin (Boddaert) [Garden Warbler]

   Sylvia communis Latham [Whitethroat]
   Phylloscopus collybita (Viellot) [Chiffchaff]
   Phylloscopus trochilus (L.) [Willow Warbler]
   Regulus regulus (L.) [Goldcrest]
  Muscicapidae (Flycatchers)
   Musciapa striata (Pallas) [Spotted Flycatcher]
  Aegithalidae (Long-tailed Tits and Allies)
   Aegithalos caudatus (L.) [Long-tailed Tit]
  Paridae (Tits)
   Parus ater L. [Coal Tit]
   Parus caeruleus L. [Blue Tit]
   Parus major L. [Great Tit]
   Parus montanus Conrad [Willow Tit]
   Parus palustris L. [Marsh Tit]
  Sittidae (Nuthatches)
   Sitta europaea L. [Nuthatch]
  Certhiidae (Treecreepers)
   Certhia familiaris L. [Treecreeper]
  Corvidae (Crows and Allies)
   Garrulus glandarius (L.) [Jay]
   Pica pica (L.) [Magpie]
   Pyrrhocorax pyrrhocorax (L.) [Chough]
   Corvus corax L. [Raven]
   Corvus corone L. [Carrion Crow]
   Corvus frugilegus L. [Rook]
   Corvus monedula L. [Jackdaw]
  Sturnidae (Starlings)
   Sturnus vulgaris L. [Starling]
  Passeridae (Sparrows and Allies)
   Passer domesticus (L.) [House Sparrow]
   Passer montanus (L.) [Tree Sparrow]
  Fringillidae (Finches)
   Fringilla coelebs L. [Chaffinch]
   Carduelis cannabina (L.) [Linnet]
   Carduelis carduelis (L.) [Goldfinch]
   Carduelis chloris (L.) [Greenfinch]
   Carduelis flammea (L.) [Redpoll]
   Pyrrhula pyrrhula (L.) [Bullfinch]
  Emberizidae (Buntings)
   Emberiza citrinella L. [Yellow Hammer]
   Emberiza schoeniclus (L.) [Reed Bunting]
MAMMALIA (Mammals)
  Insectivora (Insectivores)
  Erinaceida
   Erinaceus europaeus L. [Hedgehog]
  Talpidae
   Talpa europaea L. [Mole]
  Soricidae
   Sorex araneus L. [Common Shrew]
   Neomys fodiens (Pennant) [Water Shrew]

Chiroptera (Bats)

Rhinolophidae

Rhinolophus ferrumequinum (Schreber) [Greater Horseshoe Bat]

Rhinolophus hipposideros (Bechstein) [Lesser Horseshoe Bat]

Vespertilionidae

Myotis bechsteinii (Kuhl) [Bechstein's Bat]

Myotis mystacinus (Kuhl) [Whiskered Bat]

Myotis nattereri (Kuhl) [Natterer's Bat]

Myotis daubentoni (Kuhl) [Daubenton's Bat]

Eptesicus serotinus (Schreber) [Serotine Bat]

Pipistrellus pipistrellus (Shreber) [Pipistrelle Bat]

Barbastella barbastellus (Schreber) [Barbastelle Bat]

Plecotus auritus (L.) [Brown Long-eared Bat]

Lagomorpha (Lagomorphs)

Leporidae (Rabbits and Hares)

Oryctolagus cuniculus (L.) [Rabbit]

Rodentia (Rodents)

Sciuridae (Squirrels)

Sciurus carolinensis Gmelin [Grey Squirrel]

Muridae (Voles, Rats and Mice)

Clethrionomys glareolus (Schreber) [Bank Vole]

Microtus agrestis (L.) [Field Vole]

Avicola terrestris (L.) [Water Vole]

Apodemus sylvaticus (L.) [Wood Mouse]

Mus domesticus (L.) [House Mouse]

Rattus norvegicus (Berkenhout) [Common Rat]

Rattus rattus (L.) [Ship Rat]

Carnivora (Terrestrial Carnivores)

Canidae (Dogs)

Canis familiaris L. [Feral Dog]

Vulpes vulpes (L.) [Fox]

Mustelidae (Weasels and Allies)

Mustela vison Schreber [Mink]

Meles meles (L.) [Badger]

Felidae (Cats)

Felis catus L. [Feral Cat]

Artiodactyla

Cervidae (Deer)

Cervus elaphus L. [Red Deer]

Cervus nippon Temminck [Sika Deer]

Dama dama (L.) [Fallow Deer]

Capreolus capreolus (L.) [Roe Deer]

Muntiacus reevesi (Ogilby) [Muntjac]

Hydropotes inermis Swinhoe [Chinese Water Deer]

FURTHER READING

•

The list below gives some of the major texts which assist in the study of the habitats discussed in this book. They are listed under sections dealing with practical techniques (including the identification of plants and animals), practical conservation, new initiatives and journals.

PRACTICAL TECHNIQUES

Several texts help with the design of experiments and methods of approaching a particular ecological problem:

Chalmers, N. and Parker, P. (1989) *The Open University Project Guide*. Second edition. Field Studies Occasional Publications no. 9.

Gilbertson, D. D., Kent, M. and Pyatt, F. B. (1985) *Practical Ecology for Geography and Biology*. Unwin Hyman, London.

Williams, G. (1987) *Techniques and Field Work in Ecology*. Bell and Hyman Ltd., London.

Data Analysis

It is important to incorporate statistical techniques into the experimental design, since a poorly designed experiment can be difficult to interpret. There are a number of reference texts, but most are quite heavy going. The following are a few of the more user-friendly student texts:

Chalmers, N. and Parker, P. (1989) *The Open University Project Guide*. Second edition. Field Studies Occasional Publications no. 9.

Fowler, J. and Cohen, L. (1990) *Practical Statistics for Field Biology*. Open University Press, Milton Keynes.

Watt, T. A. (1993) *Introductory Statistics for Biology Students*. Chapman and Hall, London.

Identification

There are two types of identification guide. Some are descriptive and usually contain colour illustrations. Care should be taken when using descriptive guides since it is easy to confuse superficially similar species. Better are those texts that incorporate keys (where organisms are sequentially separated out using diagnostic characters).

Field guides are accessible descriptive guides to either specific groups of plants or animals, or to particular habitats. Collins and Countrylife publish mainly descriptive guides, while Warne keys to wildflowers, birds and trees provide an alternative which contain identification keys. Field Studies Council AIDGAP keys are user-friendly and cover a wide range of plant and animal groups, especially invertebrates. Naturalists' Handbooks (Richmond Publishing Company) are other user-friendly keys to either specific groups of invertebrate animals (mainly insects), or to the occupants of particular habitats. Other identification texts exist which are not part of a series (e.g. Skinner, 1984; Marshall and Haes, 1988).

More specialist keys are aimed towards the professional and, although they may be difficult for the beginner to use, they are usually more complete than the examples given above. Specialist keys include those produced by the Freshwater Biological Association for the identification of British freshwater invertebrates, the Linnean Society Synopses of the British Fauna covering a large number of invertebrate groups (e.g. earthworms, harvestmen, woodlice, millipedes), and the Royal Entomological Society of London handbooks for the identification of British insects. Other texts such as Stace (1997) for identifying plants do not form part of a series.

PRACTICAL CONSERVATION

Baines, C. and Smart, J. (1991) *A Guide to Habitat Creation*. Ecology Handbook no. 2. Second edition, London Ecology Unit, London.

Baines, C. (1995) Urban Areas. In Sutherland, W. J. and Hill, D. A. (editors) *Managing Habitats for Conservation*. Cambridge University Press, Cambridge.

Carr, S. and Lane, A. (1993) *Practical Conservation: Urban Habitats*. Hodder and Stoughton, Kent.

Emery, M. (1986) *Promoting Nature in Cities and Towns*. Croom Helm, London.

Gilbert, O. L. and Anderson, P. (1998) *Habitat Creation and Repair*. Oxford University Press, Oxford.

Parker, D. M. (1995) *Habitat Creation – A Critical Guide*. English Nature Science, no. 21. English Nature, Peterborough.

Tait, J., Lane, A. and Carr, S. (1988) *Practical Conservation: Site Assessment and Management Planning. Hodder and Stoughton, Kent.*

The British Trust for Conservation Volunteers produce a number of practical conservation handbooks covering woodlands, hedges, paths, drystone walls and wetlands.

NEW INITIATIVES

These can often be found through magazines such as *Urban Wildlife News* (published by English Nature). Local Authority Nature Conservation Development Plans, Strategies and Biodiversity Strategies are often available from the Planning Department of the Authority concerned. Organisations such as Groundwork and the Community Forests also have information available regarding their current projects, as do local naturalist's trusts and the organisations responsible for the management of local greensites. Much of this information provides reference to interesting urban areas. The London Ecology Unit publishes a series of handbooks covering particular habitats or areas in London. Other useful books on sites of interest in urban areas include:

Smyth, B. (1987) *City Wildspace*. Hilary Shipman, London.

Smyth, B. (1990) *The Green Guide to Urban Wildlife*. A. and C. Black, London.

JOURNALS

Several journals cover ecology, management and conservation issues including those related to urban areas, for example:

Biological Conservation;
British Wildlife;
Environmental Management;
Journal of Applied Ecology
Journal of Environmental Management;
The Journal of Practical Ecology and Conservation;
Land Contamination and Reclamation;
Restoration Ecology.

English Nature produce a newsletter (*Urban Wildlife News*) which gives updates on current initiatives in urban ecology and urban site management. They also produce a magazine (*Enact*) which covers a variety of habitat management issues, techniques and case studies.

Local natural history journals give details of surveys and sites of local importance, for example:

Essex Naturalist;
Lancashire Wildlife Journal;
The London Naturalist;
The Naturalist (North of England);
North West Naturalist;
Proceedings of the Bristol Naturalists' Society;
Sorby Record (covering Sheffield).

REFERENCES

•

Abbott, R. T. and Boss, K. J. (1989) *A Classification of the Living Mollusca*. American Malacologists Inc., Florida.

Aitchison, J. W. (1990) The commons and wastes of England and Wales 1959–1989. *Area* 22: 272–277.

Allen, J. R., Wilkinson, S. B. and Hawkins, S. J. (1995) Redeveloped docks as artificial lagoons: The development of brackish-water communities and potential for conservation of lagoonal species. *Aquatic Conservation: Marine and Freshwater Ecosystems* 5: 299–309.

Andrews, J. (1989) British estuaries: an internationally important habitat for birds. *British Wildlife* 1(2): 76–88.

Andrews, J. (1991) Principles of restoration of gravel pits for wildlife. *British Wildlife* 2(2): 80–88.

Arnold, E. N. and Burton, J. A. (1978) *Reptiles and Amphibians*. Collins, London.

Ash, H. J. (1991) Soils and vegetation in urban areas. In: Bullock, P. and Gregory, P. J. (eds) *Soils in the urban environment*. Blackwell Scientific Publications, Oxford, pp. 153–170.

Bailey, D. E. (1995) *Habitat Reconstruction as a Technique for the Reclamation of Limestone Quarry Faces*. Unpublished PhD thesis, Manchester Metropolitan University.

Bailey, D. E., Gagen, P. J. and Gunn, J. (1992) *The Revegetation of Landforms Constructed by Restoration Blasting*. Working Paper 92/01, Manchester Polytechnic.

Baines, C. (1995) Urban Areas. In: Sutherland, W. J. and Hill, D. A. (eds) *Managing Habitats for Conservation*. Cambridge University Press, Cambridge, pp. 362–380.

Baines, C. and Smart, J. (1991). *A Guide to Habitat Creation*. Ecology Handbook no. 2, second edition, London Ecology Unit, London.

Barker, G. and Graf, A. (1989) *Principles for Nature Conservation in Towns and Cities*. Urban Wildlife Now no. 3, Nature Conservancy Council, Peterborough.

Baur, B. and Baur, A. (1993) Climatic warming due to thermal radiation from an urban area as a possible cause for the local extinction of a land snail. *Journal of Applied Ecology* 30: 333–340.

Becker, D. (1988) *Control and Removal of 'Rhododendron ponticum' on RSPB Reserves in England and Wales*. The Royal Society for the Protection of Birds, Sandy, Bedfordshire.

Beerling, D. J. and Perrins, J. M. (1993) Biological flora of the British Isles no. 177: *Impatiens glandulifera* Royle. *Journal of Ecology* 81: 367–382.

Beerling, D. J., Bailey, J. P. and Conolly, A. P. (1994) Biological flora of the British Isles no. 183: *Fallopia japonica* (Houtt.) Ronse Decraene. *Journal of Ecology* 82: 959–979.

Begon, M., Harper, J. L. and Townsend, C. R. (1996) *Ecology: Individuals, Populations and Communities*. Third edition. Blackwell Science Ltd., Oxford.

Bell, W. J. (1981) *The Laboratory Cockroach*. Chapman and Hall, London.

Benton, T. G. (1992) The ecology of the scorpion *Euscorpius flavicaudis* in England. *Journal of Zoology, London* 226: 351–368.

Beresford, A. K. C. (1995) Redevelopment of the port of Cardiff. *Ocean and Coastal Management* 27(1–2): 93–107.

Beyer, L., Blume, H-P., Elsner, D-C. and Willnow, A. (1995) Soil organic matter composition and microbial activity in urban soils. *The Science of the Total Environment* 168: 267–278.

Birkhead, T. (1991) *The Magpies*. T. and A. D. Poyser, London.

Bloxham, M. G. (ed.) (1986) *Wildlife of the Sandwell Valley*. Sandwell Valley Field Naturalists' Club, West Midlands.

Bower, J. S., Broughton, G. F. J., Willis, P. G. and Clark, H. (1995) *Air Pollution in the UK: 1993/4*.

AEA Technology: National Environmental Technology Centre, Abingdon, Oxfordshire.

Bowman, I. (1991) History. In: Carter, P. (ed.) *The Forth and Clyde Canal Guidebook*. Forth and Clyde Canal Society, Glasgow.

Bradshaw, A. D. (1989) Wasteland management and restoration in western Europe. *Journal of Applied Ecology* 26: 775–786.

Bradshaw, A. D. and Chadwick, M. J. (1980) *The Restoration of Land*. Blackwell Scientific Publications, Oxford.

Briggs, J. (1996) Canals – wildlife value and restoration issues. *British Wildlife* 7(6): 365–377.

Brix, H. (1994) Use of constructed wetlands in water pollution control: historical development, present status and future perspectives. *Wat. Sci. Tech.* 30(8): 209–223.

Brown, A. (1992) *The UK Environment*. The Department of the Environment, Government Statistical Service, HMSO, London.

Burman, P. and Stapleton, H. (1988) *The Churchyards Handbook*. Third edition. Church House Publishing, London.

Calhoon, R. E. and Haspel, C. (1989) Urban cat populations compared by season, subhabitat and supplemental feeding. *Journal of Animal Ecology* 58: 321–328.

Cannon, P. F. and Minter, D. W. (1984) *Rhytisma Acerinum. CMI Descriptions of Pathogenic Fungi and Bacteria* Number 791. Commonwealth Mycological Institute, Surrey.

Carter, D. J. and Hargreaves, B. (1986) *Caterpillars of Butterflies and Moths in Britain and Europe*. Collins, London.

Carter, P. (ed.) (1991) *The Forth and Clyde Canal Guidebook*. Forth and Clyde Canal Society, Glasgow.

Chandler, T. J. (1976) The climate of towns. In: Chandler, T. J. and Gregory, S. (eds) *The climate of the British Isles*. Longman, London, pp. 307–329.

Chatters, C. (1996) Conserving rare plants in muddy places. *British Wildlife* 7(5): 281–286.

City of Cardiff (1995) *Nature Conservation Strategy*. Part 1. Cardiff.

Clark, N. A. and Prys-Jones, R. P. (1994) Low tide distribution of wintering waders and shelduck on the Severn Estuary in relation to the proposed tidal barrage. *Biological Journal of the Linnean Society* 51: 199–217.

Clarkson, K. and Garland, S. (1988) Colonisation of Sheffield's urban wastelands – vascular plants. In: Whiteley, D. (ed.) *Sheffield's urban wildlife. Sorby Record* 25: 5–21.

Clemens, J., Bradley, C. and Gilbert, O. L. (1984) Early development of vegetation on urban demolition sites in Sheffield, England. *Urban Ecology* 8: 139–148.

Clinging, V. and Whiteley, D. (1980) *Mammals of the Sheffield Area*. Sorby Record Special Series 3. Sorby Natural History Society and Sheffield City Museums, Sheffield.

Corbet, G. B. and Harris, S. (1991) (eds) *The Handbook of British Mammals*. Third Edition. Blackwell Science Ltd., Oxford.

Corke, D. (1991) Stinging nettle butterflies. *British Wildlife* 2(6): 325–334.

Cousins, S. H. (1982) Species size distributions of birds and snails in an urban area. In: Bornkamm, R., Lee, J. A. and Seaward, M. R. D. (eds) *Urban Ecology*. Proceedings of the Second European Ecological Symposium, Berlin. Blackwell Scientific Publications, Oxford. pp. 99–109.

Cramp, S. (1977–1994) *Handbook of the Birds of Europe, the Middle East and North Africa: The Birds of the Western Palearctic*. Volumes 1, 2, 3, 4, 5 and 8. Oxford University Press, Oxford.

Crawley, M. J. (1987) What makes a community invasible? In: Gray, A. J., Crawley, M. J. and Edwards, P. J. (eds) *Colonization, Succession and Stability*. Symposium of the British Ecological Society 26. Blackwell Scientific Publications, Oxford.

Cresswell, P., Harris, S. and Jeffries, D. J. (1990) *The History, Distribution, Status and Habitat Requirements of the Badger in Britain*. Nature Conservancy Council, Peterborough.

Cullen, W. R. (1995) *The Colonisation and Establishment of Surface Active Invertebrate Communities in Restoration Blasted Limestone Quarries*. Unpublished PhD Thesis. The Manchester Metropolitan University.

Cullen, W. R., Wheater, C. P. and Dunleavy, P. D. (1998) Establishment of species-rich vegetation on reclaimed limestone quarry faces in Derbyshire, UK. *Biological Conservation* 84: 25–33.

Czechowski, W. (1980) Influence of the manner of managing park areas and their situation on the formation of the communities of carabid beetles

(Coleoptera, Carabidae). *Fragmenta Faunistica* 25(12): 199–217.

Darlington, A. (1969) *Ecology of Refuse Tips.* Heinemann, London.

Darlington, A. (1981) *Ecology of Walls.* Heinemann, London.

Darolová, A. (1992) Nesting of *Falco tinnunculus* (Linnaeus, 1758) in the urban agglomeration of Bratislava. *Biológia (Bratislava)* 47(5): 389–397.

Dautel, H., Kahl, O., Scheurer, S. and Knülle, W. (1994) Seasonal activity of the pigeon tick *Argus reflexus* (Acari: Argasidae) in Berlin, Germany. *Folia Parasitologica* 41: 155–160.

Davies, R. (1991) Renewal. In: Carter, P. (ed.) *The Forth and Clyde Canal Guidebook.* Forth and Clyde Canal Society, Glasgow.

Davis, B. N. K. (1976) Wildlife, urbanisation and industry. *Biological Conservation* 10: 249–291.

Davis, B. N. K. (1979) Chalk and limestone quarries as wildlife habitats. *Minerals and the Environment.* 1: 48–56.

Davis, B. N. K. (1982a) Habitat diversity and invertebrates in urban areas. In: Bornkamm, R., Lee, J. A. and Seaward, M. R. D. (eds) *Urban Ecology.* Proceedings of the Second European Ecological Symposium, Berlin. Blackwell Scientific Publications, Oxford, pp. 49–63.

Davis, B. N. K. (ed.) (1982b) Regional variation in quarries. In: Davis, B. N. K. (ed.) *Ecology of Quarries.* Institute of Terrestrial Ecology, Cambridge.

Davis, B. N. K. (1986) Colonization of newly created habitats by plants and animals. *Journal of Environmental Management* 22: 361–371.

Davis, B. N. K. (1991) *Insects on Nettles.* Second edition. Naturalists' Handbooks 1, Richmond Publishing Co. Ltd., Slough.

Davis, B. N. K. and Jones, P. E. (1978) The ground arthropods of some chalk and limestone quarries in England. *Journal of Biogeography* 5: 159–171.

Davis, B. N. K., Lakhani, K. H. and Brown, M. C. (1993) Experiments on the effects of fertilizer and rabbit grazing treatments upon the vegetation of a limestone quarry floor. *Journal of Applied Ecology* 30: 615–628.

Dennis, E. (1993) The living churchyard – sanctuaries for wildlife. *British Wildlife* 4: 230–241.

Department of the Environment (1981) *Wildlife and Countryside Act.* HMSO, London.

Department of the Environment (1989) *The Reclamation of Mineral Workings.* DoE and Welsh Office Circular MPG7, HMSO, London.

Dickman, C. R. (1987) Habitat fragmentation and vertebrate species richness in an urban environment. *Journal of Applied Ecology* 24: 337–351.

Dickman, C. R. and Doncaster, C. P. (1987) The ecology of small mammals in urban habitats. I. Populations in a patchy environment. *Journal of Animal Ecology* 56: 629–640.

Dobson, F. S. (1992) *Lichens.* Richmond Publishing Co. Ltd., Slough.

Dochinger, L. S. (1980) Interception of airborne particles by tree plantings. *Journal of Environmental Quality* 9(2): 265–268.

Dochinger, L. S. (1988) Air pollution impacts on forest trees: foliar response. *Perspectives in Environmental Botany* 2: 1–24.

Don, B. A. C. (1985) The use of drey counts to estimate Grey squirrel populations. *Journal of Zoology, London* 206: 282–286.

Doncaster, C. P. (1994) Factors regulating local variations in abundance: field tests on hedgehogs *Erinaceus europaeus*. *Oikos* 69: 182–192.

Doncaster, C. P. and Macdonald, D. W. (1992) Optimum group size for defending heterogenous distributions of resources: a model applied to Red Foxes, *Vulpes vulpes*, in Oxford City. *J. Theor. Biol.* 159: 189–198.

Doody, J. P. (1992) Coastal zone conservation. In: Duncan, K. A., Kaznowska, S. and d'A Laffoley, D. (eds) Number 5, *Marine Nature Conservation in England – Challenge and Prospects.* Proceedings of a one day seminar hosted by English Nature and the Marine Forum, English Nature, pp. 3–11.

Dowdeswell, W. H. (1987) *Hedgerows and Verges.* Allen and Unwin, London.

Eaton, J. W. (1989) Ecological aspects of water management in Britain. *Journal of Applied Ecology* 26: 835–849.

Edlin, H. L. and Mitchell, A. F. (1985) *Broadleaves.* Forestry Commission Booklet no. 20, HMSO, London.

Edwards, C. A. and Bohlen, P. J. (1996) *Biology and Ecology of Earthworms.* Chapman and Hall, London.

Edwards, R. and Howell, R. (1989) Welsh rivers and reservoirs: management for wildlife conservation. *Regulated Rivers: Research and Management* 4: 213–223.

Eggo, N. (1990) *SSSIs in Urban Areas in England.* Nature Conservancy Council, Peterborough.

Emery, M. (1986) *Promoting Nature in Cities and Towns.* Croom Helm, London.

Emmet, A. M. and Heath, J. (1989–1991) *The Moths and Butterflies of Great Britain and Ireland.* Volume 7, parts 1 and 2. Harley Books, Essex.

Eversham, B. C., Roy, D. B. and Telfer, M. G. (1996) Urban, industrial and other manmade sites as analogues of natural habitats for Carabidae. *Ann. Zool. Fennici* 33: 149–156.

Eversham, B. C. and Telfer, M. G. (1994) Conservation value of road-side verges for stenotopic heathland Carabidae: corridors or refugia? *Biodiversity and Conservation* 3: 538–545.

Ewins, P. J. and Bazely, D. R. (1995) Phenology and breeding success of feral rock doves, *Columba livia*, in Toronto, Ontario. *Canadian Field-Naturalist* 109(4): 426–432.

Eyre, M. D. and Luff, M. L. (1995) Coleoptera on post-industrial land: a conservation problem? In: Proceedings of the British Ecological Society Conference: *Recent Advances in Urban and Post-Industrial Wildlife Conservation and Habitat Creation.* University of Leicester. Land Contamination and Reclamation 3(2): 132–134.

Fail, J. Jr. (1995) Teaching ecology in urban environments. *The American Biology Teacher* 57(8): 522–525.

Ferns, P. N., Hastings, M. P. and Shaw, T. L. (1984) Minimizing the possible effects of a tidal power barrage on the shorebird populations of the Severn estuary. *Journal of Environmental Management* 18: 131–143.

Figley, W. K. and VanDruff, L. W. (1982) The ecology of urban mallards. *Wildlife Monographs* 81: 6–39.

Fisk, E. J. (1978) The growing use of roofs by nesting birds. *Bird-Banding* 49: 134–141.

Fitzgibbon, C. D. (1993) The distribution of grey squirrel dreys in farm woodland: the influence of wood area, isolation and management. *Journal of Applied Ecology* 30: 736–742.

Foelix, R. F. (1982) *Biology of Spiders.* Harvard University Press, London.

Ford, H. A. (1987) Bird communities on habitat islands in England. *Bird Study* 34: 205–218.

Forsythe, T. G. (1987) *Common Ground Beetles.* Naturalists' Handbooks 8, Richmond Publishing Co. Ltd., Slough.

Forth and Clyde Canal Joint Advisory Committee (1995) *Lowland Canals Sustainable Development Strategy.* Forth and Clyde Canal Nature Conservation Strategy, Glasgow.

Frazer, D. (1983) *Reptiles and Amphibians.* Collins, London.

Fricker, C. R. (1984) A note on salmonella excretion in the blackheaded gull (*Larus ribibundus*) feeding at sewage-treatment works. *Journal of Applied Bacteriology* 56(3): 499–502.

Fuller, R. J. and Glue, D. E. (1981) The impact on bird communities of the modernisation of sewage treatment works. *Effluent and Water Treatment Journal* 21: 27–31.

Fuller, R. J. and Peterken, G. F. (1995) Woodland and scrub. In: Sutherland, W. J. and Hill, D. A. (eds) *Managing Habitats for Conservation.* Cambridge University Press, Cambridge, pp. 327–361.

Furness, R. W. and Monaghan, P. (1987) *Seabird Ecology.* Blackie, Glasgow.

Gagen, P. J., Bailey, D. E. and Gunn, J. (1992) *Landform Construction by Restoration Blasting.* Working Paper 92/02, The University of Huddersfield.

Gemmell, R. P. (1977) *Colonization of Industrial Wasteland.* Studies in Biology 80, Edward Arnold, London.

Gilbert, O. L. (1981) Plant communities in an urban environment. *Landscape Res.* 6: 5–7.

Gilbert, O. L. (1988) Urban demolition sites: A neglected habitat. *Bull. Brit. Lich. Soc.* 62: 1–3.

Gilbert, O. L. (1989) *The Ecology of Urban Habitats.* Chapman and Hall, London.

Gilbert, O. L. (1992a). *The Flowering of Cities: the Natural Flora of Urban 'Commons'.* English Nature, Peterborough.

Gilbert, O. L. (1992b) *Rooted in Stone: the Natural Flora of Urban walls.* English Nature, Peterborough.

Gilbert, O. L. (1992c) The ecology of an urban river. *British Wildlife* 3(3): 129–136.

Gilbert, O. L. (1994) Japanese Knotweed – what problem? *Urban Wildlife News* 11(3): 1–2.

Gilbert, O. L. and Anderson P. (1998) *Habitat Creation and Repair.* Oxford University Press, Oxford.

Gill, J. P., Townsley M. and Mudge G. P. (1996) *Review of the Impacts of Wind Farms and Other Aerial Structures Upon Birds.* Scottish Natural

Heritage Review No. 21, Scottish Natural Heritage, Battleby, Perth.

Glue, D. and Muirhead, L. (1991) Garden bird studies. In: Stroud, D. and Glue, D. (eds). *Britain's Birds in 1989–90: the Conservation and Monitoring Review.* British Trust for Ornithology / Nature Conservancy Council, Thetford.

Goldstein-Golding, E. L. (1991) The ecology and structure of urban greenspaces. In Bell, S. S., McCoy, E. D. and Mushinsky, H. R. (eds) *Habitat Structure.* Chapman and Hall, London.

Goszczynyński, J., Jabloński, P., Lesiński, G. and Romanowski, J. (1993) Variation in the diet of tawny owl *Strix aluco* L. along an urbanization gradient. *Acta Ornithologica* 27(2): 113–123.

Gray, A. (ed.) (1992) *The Ecological Impact of Estuarine Barrages.* British Ecological Society Ecological Issues No. 3. Field Studies Council, Shrewsbury.

Greenwood, E. F. and Gemmell, R. P. (1978) Derelict industrial land as a habitat for rare plants in S. Lancs. (v.c. 59) and W. Lancs. (v.c. 60). *Watsonia* 12: 33–40.

Grime, J. P., Hodgson, J. G. and Hunt, R. (1988) *Comparative Plant Ecology.* Unwin Hyman, London.

Groundwork Trust (1990) *Sutton Bank Mill Wildlife Park Management Plan.* St Helens, Knowsley and Sefton Groundwork Trust, Merseyside.

Gurnell, J. (1987) *The Natural History of Squirrels.* Christopher Helm, London.

Harding, P. T. and Sutton, S. L. (1985) *Woodlice in Britain and Ireland: Distribution and Habitat.* Natural Environment Research Council and the Institute of Terrestrial Ecology, Huntingdon.

Harris, E. and Harris, J. (1991) *Wildlife Conservation in Managed Woodlands and Forests.* Basil Blackwell, Oxford.

Harris, J. A., Birch, P. and Palmer, J. (1996) *Land Restoration and Reclamation: Principles and Practice.* Longman, Essex.

Harris, S. (1981) An estimation of the number of foxes (*Vulpes vulpes*) in the city of Bristol, and some possible factors affecting their distribution. *Journal of Applied Ecology* 18: 455–465.

Harris, S. (1984) Ecology of urban badgers *Meles meles*: distribution in Britain and habitat selection, persecution, food and damage in the city of Bristol. *Biological Conservation* 28: 349–375.

Harris, S. and Cresswell, W. J. (1987) Dynamics of a suburban badger (*Meles meles*) population. *Symp. Zool. Soc. Lond.* 58: 295–311.

Harris, S. and Rayner, J. M. V. (1986) A discriminant function analysis of the current distribution of urban foxes (*Vulpes vulpes*) in Britain. *Journal of Animal Ecology* 55: 605–611.

Harris, S., Morris, P., Wray, S. and Yalden, D. (1995) *A Review of British Mammals: Population Estimates and Conservation Status of British Mammals other than Cetaceans.* Joint Nature Conservation Council, Peterborough.

Harrison, C., Burgess, J., Millward, A. and Dawe, G. (1995) *Accessible Natural Greenspace in Towns and Cities: A Review of Appropriate Size and Distance Criteria.* English Nature Reports, Number 153. English Nature, Peterborough.

Harrison, P. (1983) *Seabirds.* Croom Helm, Kent.

Hawksworth, D. L. and Rose, F. (1976) *Lichens as Pollution Monitors.* Studies in Biology 66. Edward Arnold, London.

Hayward, P. J. and Ryland, J. S. (1990) *The Marine Fauna of the British Isles and North-west Europe.* Clarendon Press, Oxford.

Hellawell, J. M. (1986) *Biological Indicators of Freshwater Pollution and Environmental Management.* Elsevier Applied Science Publishers, London.

Hengeveld, H. and de Vocht, C. (1980/81) Role of water in urban ecology. *Special Issue of Urban Ecology* 6: i–viii + 1–362.

Hextell, T. and Hackett, P. (1994) *Birds of Sandwell Valley 1993.* RSPB, West Midlands.

Hibberd, B. G. (1989) *Urban Forestry Practice.* Forestry Commission Handbook 5, HMSO, London.

Hill, M. O., Preston, C. D. and Smith, A. J. E. (1991–94) *Atlas of the Bryophytes of Britain and Ireland.* Volumes 1–3. Harley Books, Colchester.

Hodge, S. (1993) *Making It Work.* Landscape Design 217: 42–45.

Hodgson, J. G. (1982) The botanical interest and value of quarries. In: Davis, B. N. K. (ed.) *Ecology of Quarries.* Institute of Terrestrial Ecology, Cambridge.

Hodson, N. L. and Snow, D. W. (1965) The road deaths enquiry, 1960–61. *Bird Study.* 12: 90–99.

Holland, C. C., Honea, J., Gwin, S. E. and Kentula, M. E. (1995) Wetland degradation and loss in the rapidly urbanizing area of Portland, Oregon. *Wetlands* 15(4): 336–345.

Holland, P. K, Sutton, J. T. and Spence, I. M. (1984) *Breeding Birds in Greater Manchester*. Manchester Ornithological Society, Manchester.

Hopkin, S. P. (1991) A key to the woodlice of Britain and Ireland. *Field Studies* 7: 599–650.

Hopkin, S. P. and Martin, M. H. (1985) Assimilation of zinc, cadmium, lead, copper, and iron by the spider *Dysdera crocata*, a predator of woodlice. *Bulletin of Environmental Contamination and Toxicology* 34:183–187.

Hopkin, S. P., Hardisty, G. N. and Martin, M. H. (1986) The woodlouse *Porcellio scaber* as a 'biological indicator' of zinc, cadmium, lead and copper pollution. *Environmental Pollution Series B* 11: 271–290.

Hounsome, M. V. (1979) Bird life in the city. In: Laurie, I. (ed.) *Nature in Cities*. Wiley, Chichester.

Humphries, R. N. (1979) Landscaping hard quarry rock faces. *Landscape Design* 127: 34–37.

Humphries, R. N. (1980) The development of wildlife interest in limestone quarries. *Reclamation Review*. 3: 197–207.

Jansen, J., van Knapen, F., Schreurs, M and van Wijngaarden, Th. (1993) Toxocara eggs in public parks and sand-boxes in Utrecht. *Tijdschrift voor Diergeneeskunde* 118(19): 611–614.

Jim, C. Y. (1992) Tree-habitat relationships in urban Hong Kong. *Environmental Conservation* 19(3): 209–218.

Johnston, J. (1990) *Nature Areas For City People*. Ecology Handbook 14, London Ecology Unit, London.

Jones, A. (1991) British wildlife and the Law: a review of the species protection provisions of the Wildlife and Countryside Act 1981. *British Wildlife* 2(6): 345–358.

Jones, F. H. (1994) Barrage developments in the Welsh Region: the role of the National Rivers Authority in protecting the aquatic environment. *Journal of the Institution of Water and Environmental Management* 8: 432–439.

Jones, M. J. (1985) Distribution and activity of Common Gulls and Black-headed Gulls on grassland in Manchester. *Bird Study* 32: 104–112.

Jorde, D.G, Kropu, G. L., Crawford, R. D. and Hay, M. A. (1984) Effects of weather on habitat selection and behaviour of mallards wintering in Nebraska. *Condor* 86: 258–265.

Kellert, S. R. (1991) Japanese perceptions of wildlife. *Conservation Biology* 5(3): 297–308.

Kennedy, C. E. J. and Southwood, T. R. E. (1984) The number of species of insects associated with British trees: a re-analysis. *Journal of Animal Ecology*. 53: 455–478.

Kevan, D. K. (1945) The Coleoptera of an Edinburgh garden. *Ent. Mon. Mag.* 81: 112–113.

Kirby, J. (1995) *A Study of the Habitat Preferences of Water Voles (Arvicola terrestris) in the River Leen, Nottingham*. Unpublished MSc Thesis, The Manchester Metropolitan University.

Kloet, G. S. and Hincks, W. D. (1964–1975) *A Check List of British Insects*. Second edition. Parts 1–5. The Royal Entomological Society of London, London.

Kooiker, G. (1994) Influence of magpie *Pica pica* on urban bird populations in the city of Osnabruck, northwest Germany. *Vogelwelt* 115(1): 39–44.

Kozakiewicz, M. and Jurasińska, E. (1989) The role of habitat barriers in woodlot recolonization by small mammals. *Holarctic Ecology*. 12: 106–111.

Krebs, J. R. and Davies, N. B. (1993) *An Introduction to Behavioural Ecology*. Third edition. Blackwell Science, Oxford.

Kubečke, J. and Duncan, A. (1994) Low fish predation pressure in the London reservoirs: I. Species composition, density and biomass. *Int. Revue ges. Hydrobiol.* 79(1): 143–155.

Lack, D. (1972) *The Life of the Robin*. Collins, London.

Lagey, K., Duinslaeger, L. and Vanderkelen, A. (1995) Burns induced by plants. *Burns* 21(7): 542–543.

Lampard, D. and Morgan, C. (1991) Wildlife. In: Carter, P. (ed.) *The Forth and Clyde Canal Guidebook*. Forth and Clyde Canal Society, Glasgow.

Land Care Associates (1995) *Nature Conservation Strategy for Birmingham*. Third Draft. Birmingham City Council, Birmingham.

Laurie, I. (1979) Urban commons. In: Laurie, I. (ed.) *Nature in Cities*. Wiley, Chichester.

Leith, I. D. and Fowler, D. (1988) Urban distribution of *Rhyisma acerinum* (Pers.) Fries (tar spot) on sycamore. *New Phytologist* 108(2): 175–181.

Lever, C. (1994) *Naturalized Animals*. T. and A. D. Poyser Natural History, London.

Lewis, G. and Williams, G. (1984) *Rivers and Wildlife Handbook*. RSPB, Sandy, Bedfordshire.

Lickorish, S., Luscombe, G. and Scott, R. (1997) *Wildflowers Work*. Landlife, Liverpool.

Linde, D. T., Watschke, T. L., Jarrett, A. R. and Borger, J. A. (1995) Surface runoff assessment from creeping bentgrass and perennial ryegrass turf. *Agronomy Journal* 87: 176–182.

Llewellyn, P. J. and Shackley, S. E. (1996) The effects of mechanical beach-cleaning on invertebrate populations. *British Wildlife* 7(3): 147–155.

Luniak, M. (1995) Peregrine falcon *Falco perigrinus* in cities – the background for its planned reintroduction in Warsaw. *Acta Ornithologica* 30: 53–62.

Lussenhop, J. (1977) Urban cemeteries as bird refuges. *Condor* 79: 456–461.

Mabey, R. (1996) *Flora Britannica*. Sinclair-Stevenson, London.

MacArthur, R. H. and Wilson, E. O. (1967) *The Theory of Island Biogeography*. Princeton University Press, Princeton.

McClure, H. E. (1989) What characterises an urban bird? *Journal of The Yamashina Institute for Ornithology* 21(2): 178–192.

Macdonald, D. W. (1987) *Running with the Fox*. Unwin Hyman, London.

McDonnell, M. J. and Pickett, S. T. A. (1990) Ecosystem structure and function along urban-rural gradients: an unexploited opportunity for ecology. *Ecology* 71(4): 1232–1237.

MacGregor, H. (1995) Crested newts – ancient survivors. *British Wildlife* 7(1): 1–8.

Mader, H-J., Schell, C. and Kornacker, P. (1990) Linear barriers to arthropod movements in the landscape. *Biological Conservation* 54: 209–222.

Maitland, P. S. and Lyle, A. A. (1991) Conservation of freshwater fish in the British Isles: the current status and biology of threatened species. *Aquatic Conservation: Marine and Freshwater Ecosystems* 1: 25–54.

Majdi, H. and Persson, H. (1989) Effects of road-traffic pollutants (lead and cadmium) on tree fine-roots along a motor road. *Plant and Soil* 119: 1–5.

Majerus, M. E. N. (1989) Melanic polymorphism in the peppered moth *Biston betularia*, and other Lepidoptera. *Journal of Biological Education* 23(4): 267–284.

Mao, L-S., Gao, Y. and Sun, W-Q. (1993) Influences of street tree systems on summer micro-climate and noise attenuation in Nanking City, China. *Arboricultural Journal* 17: 239–251.

Margulis, L. and Schwartz, K. V. (1988) *Five Kingdoms*. Second edition. Freeman, New York.

Marine Task Force (1994) *Important Areas for Marine Wildlife Around England*. English Nature, Peterborough.

Marsalek, J., Barnwell, T. O., Geiger, W., Grottker, M., Huber, W. C., Saul, A. J., Schilling, W. and Torno, H. C. (1993) Urban drainage systems: design and operation. *Wat. Sci. Tech.* 27(12): 31–70.

Marshall, J. A. and Haes, E. C. M. (1988) *The Grasshoppers And Allied Insects Of the British Isles*. Harley Books, Essex.

Martin, M. H. and Caughtrey, P. J. (1987) Cycling and fate of heavy metals in a contaminated woodland ecosystem. In: Caughtrey, P. J., Martin, M. H. and Unsworth, M. H. (eds) *Pollutant Transport and Fate in Ecosystems*. Special Publication of the British Ecological Society 6: 319–336.

Mersey Forest (1995) *Vision to Reality*. The Mersey Forest, Warrington.

Millward, A. and Mostyn, B. (1989) *People and Nature in Cities*. Urban Wildlife Now, Number 2 Nature Conservancy Council, Peterborough.

Milner, E. J. (1993) Spiders and disturbance on Hampstead Heath and some other London grasslands. *The London Naturalist* 72: 85–99.

Mitchell, A. (1985) *Conifers*. Forestry Commission Booklet no. 15, HMSO, London.

Mitchell-Jones, A. J. (1989) The effect of legal protection on bat conservation in Britain. In: Hanák, V., Horáček, I. and Gaisler, J. (eds) *European Bat Research Symposium*. Prague August 18–23, 1987: 671–676.

Mitchell-Jones, A. J., Cooke, A. S., Boyd, I. L. and Stebbings, R. E. (1989) Bats and remedial timber treatment chemicals – a review. *Mammal Review* 19(3): 93–110.

Morris, M. G., Thomas, J. A., Ward, L. K., Snazell, R. G., Pywell, R. F., Stevenson, M. J. and Webb, N. R. (1994) Re-creation of early-successional stages for threatened butterflies – an ecological engineering approach. *Journal of Environmental Management* 42: 119–135.

Morris, R. and Parsons, M. (1993) Dungeness – a shingle beach and its invertebrates. *British Wildlife* 4(3): 137–144.

Munguira, M. L. and Thomas, J. A. (1992) The use of road verges by butterfly and burnet populations, and the effect of roads on adult dispersal and mortality. *Journal of Applied Ecology* 29: 316–329.

Murphy, K. J. and Eaton, J. W. (1983) Effects of pleasure-boat traffic on macrophyte growth in canals. *Journal of Applied Ecology* 20: 713–729.

National Rivers Authority (1994) *Guidance for the Control of Invasive Plants Near Watercourses.* National Rivers Authority, Bristol.

Nichols, D. (1990) *Safety in Biological Fieldwork – Guidance Notes for Codes of Practice.* The Institute of Biology, London.

Nuckols, M. S. and Connor, E. F. (1995) Do trees in urban or ornamental plantings receive more damage by insects than trees in natural forests? *Ecological Entomology* 20: 253–260.

Oliver, P. G. and Meecham, C. J. (1993) *Woodlice.* Synopses of the British Fauna no. 49. The Linnean Society of London, E.J. Brill / Dr W. Backhuys, London.

Owen, J. (1991) *The Ecology of A Garden.* Cambridge University Press, Cambridge.

Oxley, D. J., Fenton, M. B. and Carmondy, G. R. (1974) The effects of roads on populations of small mammals. *Journal of Applied Ecology* 11: 51–59.

Palmer, J. and Neal, P. (1994) *The Handbook of Environmental Education.* Routledge, London.

Penfold, N. and Francis, I. (1991) Common land and nature conservation in England and Wales. *British Wildlife* 2(2): 65–76.

Peterken, G. (1993) *Woodland Conservation and Management.* Second Edition. Chapman and Hall, London.

Port, G. R. and Thompson, J. R. (1980) Outbreaks of insect herbivores on plants along motorways in the United Kingdom. *Journal of Applied Ecology* 17: 649–656.

Prach, K. (1994) Seasonal dynamics of *Impatiens glandulifera* in two riparian habitats in Central England. In: de Waal, L. C., Child, L. E., Wade, P. M. and Brock, J. H. (eds) *Ecology and Management of Invasive Riverside Plants.* John Wiley and Sons, Chichester. pp. 127–133.

Prach, K. and Pyšek, P. (1994) Clonal plants – what is their role in succession? *Folia Geobot. Phytotax., Praha* 29: 307–320.

Pycraft, D. (1994) *Japanese Knotweed: Some Notes on Control.* Horticultural Advisory Leaflet, Hort 21, The Royal Horticultural Society, Wisley.

Pye, K. and French, P. W. (1993) *Targets for Coastal Habitat Re-creation.* English Nature Science no. 13, English Nature, Peterborough.

Pyšek, P. and Prach, K. (1995) Invasion dynamics of *Impatiens glandulifera* – a century of spreading reconstructed. *Biological Conservation* 74: 41–48.

Rackham, O. (1986) *The History of the Countryside.* Dent, London.

Ratcliffe, D. A. (1974) Ecological effects of mineral exploitation in the United Kingdom and their significance to nature conservation. *Proceedings of the Royal Society of London, A* 339: 355–372.

Read, H. J. (1987) *The Effects of Heavy Metal Pollution on Woodland Leaf Litter Faunal Communities.* Unpublished PhD Thesis, University of Bristol.

Read, H. J., Wheater, C. P. and Martin, M. H. (1987) Aspects of the ecology of Carabidae (Coleoptera) from woodlands polluted by heavy metals. *Environmental Pollution* 48: 61–76.

Read, H. J., Rayner, J. M. V. and Martin, M. H. (1998) Invertebrates in woodlands polluted by heavy metals: an evaluation using canonical correspondence analysis. *Water, Air and Soil Pollution* 106: 17–42.

Reeve, N. (1994) *Hedgehogs.* T. and A. D. Poyser Natural History, London.

Reijnen, R., Foppen, R., ter Braak, C. and Thissen, J. (1995) The effects of car traffic on breeding bird populations in woodland. III. Reduction of density in relation to the proximity of main roads. *Journal of Applied Ecology* 32(1): 187–202.

Richardson, D. H. S. (1992) *Pollution Monitoring with Lichens.* Naturalists' Handbooks 19. Richmond Publishing Co. Ltd., Slough.

Roberts, M. J. (1985–1987) *The Spiders of Great Britain and Ireland.* Volumes I, II and III. Harley Books, Colchester.

Rushton, S. P., Hill, D. and Carter, S. P. (1994) The abundance of river corridor birds in relation to their habitats: a modelling approach. *Journal of Applied Ecology* 31: 313–328.

Rydell, J. (1992) Exploitation of insects around streetlamps by bats in Sweden. *Functional Ecology* 6: 744–750.

Salmon, D. G. and Fox, A. D. (1994) Changes in

the wildfowl populations wintering on the Severn Estuary. *Biological Journal of the Linnean Society* 51: 229–236.

Sanderson, R. A. (1992) Hemiptera of naturally vegetated derelict land in north-west England. *Entomologist's Gazette* 43(3): 221–226.

Sargent, C. (1984) *Britain's Railway Vegetation*. Natural Environment Research Council and Institute of Terrestrial Ecology, Cambridge.

Schaefer, M. (1982) Studies on the arthropod fauna of green urban ecosystems. In: Bornkamm, R., Lee, J. A. and Seaward, M. R. D. (eds) *Urban Ecology*. Proceedings of the Second European Ecological Symposium, Berlin. Blackwell Scientific Publications, Oxford, pp. 65–73.

Seager, J. and Abrahams, R. G. (1990) The impact of storm sewage discharges on the ecology of a small urban river. *Wat. Sci. Tech.* 22: 163–171.

Shaw, P. (1994) Orchid woods and floating islands – the ecology of fly ash. *British Wildlife* 5(3): 149–157.

Simmons, S. A. and Barker, A. (1989) *Urban Wetlands for Nature Conservation and Stormwater Control*. Urban Wildlife Now, no. 4. Nature Conservancy Council, Peterborough.

Sims, R. W. and Gerard, B. M. (1985) *Earthworms*. Synopses of the British Fauna no. 31. The Linnean Society of London, E.J. Brill, / Dr W. Backhuys, London.

Skinner, B. (1984) *Moths of the British Isles*. Viking, Middlesex.

Smith, A. J. (ed.) (1996) *Birds in Greater Manchester: Twentieth Greater Manchester Bird Report 1995*. The Greater Manchester Bird Club, Manchester.

Smith, G. C. and Carlile, N. (1993) Food and feeding ecology of breeding Silver Gulls (*Larus novaehollandiae*) in urban Australia. *Colonial Waterbirds* 16(1): 9–17.

Smyth, B. (1990) *The Green Guide to Urban Wildlife*. A. and C. Black, London.

Sodhi, N., James, P. C., Warkentin, I. G. and Oliphant, L. W. (1992) Breeding ecology of urban Merlins (*Falco columbarius*). *Canadian Journal of Zoology* 70: 1477–1483.

Sol, D. and Senar, J. C. (1995) Urban pigeon populations: stability, home range, and the effect of removing individuals. *Canadian journal of Zoology* 73: 1154–1160.

Spellerberg, I. F. and Gaywood, M. J. (1993) *Linear Features: Linear Habitats and Wildlife Corridors*. English Nature Research Reports, Number 60. English Nature, Peterborough.

Stace, C. (1997) *New Flora of the British Isles*. Second edition. Cambridge University Press, Cambridge.

Steiner, W. A. (1994) The influence of air pollution on moss-dwelling animals: 1. Methodology and the composition of flora and fauna. *Revue Suisse de Zoologie* 101(2): 533–556.

Stewart, L. A. and Dixon, A. F. G. (1989) Why big species of ladybird beetles are not melanic. *Functional Ecology* 3: 165–177.

Street, M. (1985) *The Restoration Of Gravel Pits For Wildfowl*. ARC, Chipping Sodbury.

Stroud, D. and Glue, D. (1991) *Britain's Birds in 1989–90: The Conservation and Monitoring Review*. British Trust for Ornithology / Nature Conservancy Council, Thetford.

Summers-Smith, J. D. (1988) *The Sparrows*. T. and A. D. Poyser, Calton.

Tatner, P. (1978) A review of House Martins (*Delichon urbica* L.) in part of South Manchester, 1975. *Naturalist* 103: 59–68.

Tatner, P. (1982) The breeding biology of Magpies (*Pica pica*) in an urban environment. *Journal of Zoology*, London 197: 559–581.

Thiele, H.-U. (1977) *Carabid Beetles in Their Environments*. Springer-Verlag, Berlin.

Thomas, K. (1992) A study of urban fungi in the London Borough of Harringey, 1983–1991. *London Nat.* 71: 43–60.

Tiley, G. E. D., Dodd, F. S. and Wade, P. M. (1996) Biological Flora of the British Isles No. 190: *Heracleum mantegazzianum* Sommier and Levier. *Journal of Ecology* 84: 297–319.

Tilghman, N. G. (1987) Characteristics of urban woodlands affecting breeding bird diversity and abundance. *Landscape and Urban Planning* 14: 481–495.

Tobin, B. and Taylor, B. (1996) Golf and wildlife. *British Wildlife* 7(3): 137–146.

Turnbull, D. A. and Bevan, J. R. (1995) The impact of airport de-icing on a river: the case of the Ouseburn, Newcastle upon Tyne. *Environmental Pollution* 88: 321–332.

Turner, A. (1989) *Swallows and Martins*. Christopher Helm, London.

United Nations Environment Programme (1991) *United Nations Environment Programme*

Environmental Data Report. Third edition. Basil Blackwell, Oxford.

Usher, M. B. (1977) Natural communities of plants and animals in disused quarries. In: *Papers of the Land Reclamation Conference. 1976*. Thurrock Borough Council, Grays, Essex. pp. 401–420.

Usher, M. B. (1986) Invasibility and wildlife conservation: invasive species on nature reserves. *Philosophical Transactions of the Royal Society of London, Series B* 314: 695–710.

Vermeulen, H. J. W. (1994) Corridor function of a road verge for dispersal of stenotopic heathland ground beetles (Carabidae). *Biological Conservation* 69(3): 339–349.

Village, A. (1990) *The Kestrel*. T. and A. D. Poyser, London.

Vincent, M. (1992) Conservation in estuaries. In: Duncan, K. A., Kaznowska, S. and d'A Laffoley, D. (eds) Number 5, *Marine Nature Conservation in England – Challenge and Prospects*. Proceedings of a one-day seminar hosted by English Nature and the Marine Forum, English Nature. 15–20.

Waite, M., Keech, D. and Game, M. (1993) *Nature Conservation in Camden*. Ecology Handbook 24, London Ecology Unit, London.

Wahlbrink, D. and Zucchi, H. (1994) Occurrence and settlement of carabid beetles on an urban railway embankment: a contribution to urban ecology. *Zoologische Jahrbuecher Abteilung fuer Systematik Oekologie und Geographie der Tiere* 121(2): 193–201.

Walker, S. E. and Duffield, B. S. (1983) Urban parks and open spaces – an overview. *Landscape Research* 8(2): 2–12.

Watts, S. E. J. and Smith, B. J. (1994) The contribution of highway run-off to river sediments and implications for the impounding of urban estuaries: a case study of Belfast. *The Science of the Total Environment* 146/147: 507–514.

Way, J. M. (1977) Roadside verges and conservation in Britain: A review. *Biological Conservation* 12: 65–74.

Webb, N. R. (1985) *Heathlands*. Collins, London.

Weigmann (1982) The colonization of ruderal biotypes in the City of Berlin by arthropods. In: Bornkamm, R., Lee, J. A. and Seaward, M. R. D. (eds) *Urban Ecology*. Proceedings of the Second European Ecological Symposium, Berlin. Blackwell Scientific Publications, Oxford, pp. 75–82.

Welch, S. (1986) The birds. In: Bloxham, M. G. (ed.) *Wildlife of the Sandwell Valley*. Sandwell Valley Field Naturalists' Club, West Midlands. pp. 21–35.

Wheater, C. P. and Cullen, W. R. (1997) The flora and invertebrate fauna of abandoned limestone quarries in Derbyshire, United Kingdom. *Restoration Ecology* 5(1): 77–84.

Wheater, C. P. and Read, H. J. (1996) *Animals Under Logs and Stones*. Naturalists' Handbooks 22. The Richmond Publishing Co. Ltd., Slough.

Whiteley, D. (1988) Hoverflies on urban derelict land in Sheffield. In: Whiteley D. (ed.) *Sheffield's Urban Wildlife*. *Sorby Record* 25: 45–48.

Williams, S. M. and McCrorie, R. (1990) The analysis of ecological attitudes in town and country. *Journal of Environmental Management* 31: 157–162.

Williamson, M. (1996) *Biological Invasions*. Chapman and Hall, London.

Willing, M. (1993) Land molluscs and their conservation – an introduction. *British Wildlife* 4(3): 145–153.

Wright, A. (1997) *Predicting the Distribution of Eurasian Badger* (Meles meles) *Setts*. Unpublished PhD Thesis, The Manchester Metropolitan University.

Wright, A. and Wheater, C. P. (1993) Aspects of the vegetation of an area of disused railway line in Derbyshire. *Sorby Record* 30: 57–67.

Yahner, R. H. (1988) Changes in wildlife communities near edges. *Conservation Biology* 2(4): 333–339.

Yalden, D. W. (1980a) Urban small mammals. *Journal of Zoology, London* 191: 403–406.

Yalden, D. W. (1980b) Notes on the diet of urban Kestrels. *Bird Study* 27: 235–238.

SUBJECT INDEX

●

SPECIES INDEX

●